Student Study Guide
Volume II
for
Tipler and Mosca's
Physics for Scientists and Engineers
Sixth Edition

Todd Ruskell
Colorado School of Mines

W.H. Freeman and Company
New York

Acknowledgments

I must first thank Paul Tipler and Gene Mosca for putting together an excellent textbook in the sixth edition of *Physics for Scientists and Engineers*. It has been a delight to work with. I must also thank Gene for his work on earlier versions of this study guide, which have been drawn on heavily.

I am indebted to the reviewers of this study guide: Elizabeth Behrman (Wichita State), Daniel Dale (University of Wyoming), Linnea Hess (Olympic College), and Oren Quist (South Dakota State). Their careful reading and insightful comments made this a much better study guide than it would have been otherwise. In addition, their thorough reviews helped uncover many errors that would have remained to be discovered by the users of this study guide. Their assistance is greatly appreciated. In spite of our combined best efforts, there may still be an occasional error in this study guide, and for those I assume full responsibility. Should you find errors or would like to bring another matter regarding this study guide to my attention, please do not hesitate to send them to me by using asktipler@whfreeman.com.

My wife Susan and daughter Allison provided immense support with their patience and understanding throughout the entire process of writing this study guide.

It was a pleasure to work with Susan Brennan, Clancy Marshall, Kharissia Pettus, and Kathryn Treadway who guided me through the creation of this study guide.

April, 2007

Todd Ruskell
Colorado School of Mines

Printed in the United States of America

ISBN: 1-4292-0410-9 (Volume II: Chapters 21–33)

First Printing 2008

W. H. Freeman and Company
41 Madison Avenue
New York, NY 10010
Houndmills, Basingstoke
RG21 6XS, England
www.whfreeman.com

Contents

To the Student

This study guide was written to help you master Chapters 21 through 33 of Paul Tipler and Eugene Mosca's *Physics for Scientists and Engineers*, Sixth Edition. Each chapter of the study guide is divided into sections that match the textbook, and culminates in a short quiz designed to test your mastery of the subject. Each of these sections may contain the subsections below.

In a Nutshell: A brief overview of the important concepts presented in the section. This section is designed only to remind you of the key ideas. If a concept is not clear, you should refer back to the text for more detailed explanations and derivations.

Physical Quantities and Their Units: A list of the constants, units, and physical quantities introduced in the section.

Fundamental Equations: A list of fundamental equations introduced in the section. These expressions provide underpinning for the Important Derived Results.

Important Derived Results: In many sections the Fundamental Equations are applied to specific physical situations. This application can result in important derived results that apply to only those specific situations. These results are listed here.

Common Pitfalls: Warnings about commonly-made mistakes. In addition, there are conceptual questions designed to test your understanding of the physical principles discussed in the section.

Try It Yourself: Workbook style questions following the structure of the solutions to the worked Examples in the text. Final answers, with units, are provided, but it is up to you to use the space provided to fill in the required work for the intermediate steps. Most of these questions also have a Taking It Further question designed to enhance your understanding of and ability to interpret the problem's solution. You should answer these questions in the space provided before looking at the answers in the back of the study guide.

What Is the Best Way to Study Physics?

Of course there isn't a single answer to that. It is clear, however, that you should begin early in the course to develop the methods that work best for you. The important thing is to find the system that is most comfortable and effective for you, and then stick to it.

In this course you will be introduced to numerous concepts. It is important that you take the time to be sure you understand each of them. You will have mastered a concept when you fully understand its relationships with other concepts. Some concepts will seem to contradict other concepts or even your observations of the physical world. Many of the questions in this study guide are intended to test your understanding of concepts. If you find that your understanding of an idea is incomplete, don't give up; pursue it until it becomes clear. We recommend that you keep a list of the things that you come across in your studies that you do not understand. Then, when you come to understand an idea, remove it from your list. After you complete your study of each chapter, bring your list to your most important resource, your physics instructor, and ask for assistance. If you go to your instructor with a few well-defined questions, you will very likely be able to remove any remaining items from your list.

Like the Example problems presented in the textbook, the problem solutions presented in this study guide start with basic concepts, not with formulas. We encourage you to follow this practice. Physics is a collection of interrelated basic concepts, not a seemingly infinite list of disconnected, highly specific formulas. Although at times it may seem we present long lists of formulas, do not try to memorize long lists of specific formulas and then use these formulas as the starting point for solving problems. Instead, focus on the concepts first and be sure that you understand the ideas before you apply the formulas.

Probably the most rewarding (but challenging) aspect of studying physics is learning how to apply the fundamental concepts to specific problems. At some point you are likely to think, "I understand the theory, but I just can't do the problems." If you can't do the problems, however, you probably don't understand the theory. Until the physical concepts and the mathematical equations become your tools to apply at will to specific physical situations, you haven't really learned them. There are two major aspects involved in learning to solve problems: drill and skill. By drill we mean going through a lot of problems that involve the direct application of a particular concept until you start to feel familiar with the way it applies to physical situations. Each chapter of the text contains about 35 single-concept problems for you to use as drill. Do a lot of these—at least as many as you need in order to feel comfortable handling them.

By skill we mean the ability both to recognize which concepts are involved in more advanced, multi-concept problems, and to apply those concepts to particular situations. The text has several intermediate-level and advanced-level problems that go beyond the direct application of a single concept. As you develop this skill you will master the material and become empowered. As you find that you can deal with more complex problems—even some of the advanced-level ones—you will gain confidence and enjoy applying your new skills. The examples in the textbook and the problems in this study guide are designed to provide you with a pathway from the single-concept to the intermediate-level and advanced-level problems.

A typical physics problem describes a physical situation—such as a child swinging on a swing—and asks related questions. For example: If the speed of the child is 5.0 m/s at the bottom of her arc, what is the maximum height the child will reach? Solving such problems requires you to apply the concepts of physics to the physical situation, to generate mathematical relations, and to solve for the desired quantities. The problems presented here and in your textbook are exemplars; that is, they are examples that deserve imitation. When you master the methodology presented in the worked-out examples, you should be able to solve problems about a wide variety of physical situations.

A good way to test your understanding of a specific solution is to take a sheet of paper, and—without looking at the worked-out solution of an Example problem—reproduce it. If you get stuck and need to refer to the presented solution, do so. But then take a fresh sheet of paper, start from the beginning, and reproduce the entire solution. This may seem tedious at first, but it does pay off.

This is not to suggest that you reproduce solutions by rote memorization, but that you reproduce them by drawing on your understanding of the relationships involved. By reproducing a solution in its entirety, you will verify for yourself that you have mastered a particular example problem. As you repeat this process with other examples, you will build your very own personal base of physics knowledge, a base of knowledge relating occurrences in the world around you—the physical universe—and the concepts of physics. The more complete the knowledge base that you build, the more success you will have in physics.

You should budget time to study physics on a regular, preferably daily, basis. Plan your study schedule with your course schedule in mind. One benefit of this approach is that when you study on a regular basis, more information is likely to be transferred to your long-term memory than when you are obliged to cram. Another benefit of studying on a regular basis is that you will get much more from lectures. Because you will have already studied some of the material presented, the lectures will seem more relevant to you. In fact, you should try to familiarize yourself with each chapter before it is covered in class. An effective way to do this is first to read the In a Nutshell subsections of that study guide chapter. Then thumb through the textbook chapter, reading the headings and examining the illustrations. By orienting yourself to a topic before it is covered in class, you will have created a receptive environment for encoding and storing in your memory the material you will be learning.

Another way to enhance your learning is to explain something to a fellow student. It is well known that the best way to learn something is to teach it. That is because in attempting to articulate a concept or procedure, you must first arrange the relevant ideas in a logical sequence. In addition, a dialog with another person may help you to consider things from a different perspective. After you have studied a section of a chapter, discuss the material with another student and see if you can explain what you have learned.

Chapter 21

The Electric Field I:
Discrete Charge Distributions

21.1 Charge

In a Nutshell

Electric charge is a fundamental property of matter. There are only two kinds of charge, positive and negative. We now understand that protons carry a positive charge and electrons carry an equal, but opposite, negative charge. The SI unit of charge is the **coulomb** (C), and we represent charge with the variables q or Q.

Charge is **quantized**—that is, it appears only in discrete amounts. The **fundamental unit of charge** is e. Protons carry a charge of $+e$, and electrons carry a charge of $-e$. The third component of atoms, the neutron, is electrically neutral. It has no net charge. Particles that make up protons and neutrons can have some fraction of the fundamental unit of charge. However, because this fractional amount of charge is never directly observed, it is still empirically correct to say that e is the quantum of charge.

The atomic number Z of an element tells us how many protons and electrons are present in the electrically neutral atom of that element. In general, the natural state of matter is to be electrically neutral—that is, an atom has the same number of protons and of electrons. It is possible for charge to be transferred from one object to another. In this process, typically electrons are physically ripped from one object and placed on the other.

In much the same way that we considered mass to be conserved in Volume I of the text, charge is also **conserved**. Even in circumstances that result in the creation of an electron, for instance, an equal amount of positive charge is also created so that the net charge of the universe does not change.

Physical Quantities and Their Units

Fundamental unit of charge $\qquad\qquad\qquad\qquad e = +1.602177 \times 10^{-19}$ C

Common Pitfalls

➤ We sometimes speak of the fundamental unit of charge e as "the charge of an electron," but e is a positive quantity. The charge of an electron, properly, is $-e$.

1. TRUE or FALSE: "Conservation of charge" refers to the fact that electric charge can be found only in integral multiples of the fundamental charge e.

2. Describe a set of experiments that might be used to determine if you have discovered a third type of charge other than positive and negative.

Try It Yourself #1

Your textbook claims that 10^{10} electrons can be easily transferred from fur to a rubber rod if you rub these objects together. What is the total charge transferred?

Picture: You know the number of electrons transferred, and the charge on each electron.

Solve:

Multiply the total number of transferred electrons by the charge on an electron to find the total charge transferred. Include your units and make sure the units of the result are appropriate.	
	$Q = 1.602 \times 10^{-9}$ C

Check: The units are correct.

Taking It Further: This is a pretty small number. The coulomb is actually quite a large charge. You will see in Chapter 23 why, if you happen to find 1 C of net charge in a small region of space, you generally want to treat it with respect.

Try It Yourself #2

Approximately what is the positive charge in your body due to the protons of the molecules present in your body?

Picture: You need to make a few assumptions. Humans are about 70 percent water by mass. To make life easier for yourself, assume all of your mass is water. Determine how many molecules of water are required to achieve that mass. There are 10 protons in every water molecule.

Solve:

Estimate your mass. Don't forget the units.	
Knowing that the molar mass of water is 18.0 g/mol, estimate the number of molecules of water in your body.	
Knowing there are 10 protons in each water molecule, estimate the total number of protons in your body.	

Knowing the number of protons, determine the total positive charge in your body.	
	$Q \sim 10^9$ C

Check: This is a large amount of charge. But this charge is not the net charge. The net charge of our bodies is approximately zero because each atom has an equal number of protons and electrons.

Taking It Further: From this example you can see why we often treat charge as a continuous quantity for macroscopic objects. We cannot measure in any meaningful way the loss or addition of a single electron or proton among $\sim 10^{28}$ total charges. The fractional change is too small.

21.2 Conductors and Insulators

In a Nutshell

In many materials, such as copper and other metals, some of the electrons are free to move about the entire material. Such materials are called **conductors**. The number of free electrons in a metal depends on the metal, but typically there is about one per atom.

In other materials, such as wood or glass, all the electrons are bound to nearby atoms. These materials are called **insulators**. Even when additional charge is added to an insulating material the relatively tight electronic bonds keep charge from moving freely throughout the material.

As mentioned in Section 21.1, objects can become charged by physically ripping electrons from one material and depositing those electrons on another material. **Induction** can also be used to create permanently charged conducting objects. If a charged object like a positively charged glass rod is brought near two conducting spheres, electrons will be attracted to the rod and accumulate on the closer sphere, leaving a deficit of electrons on the far sphere. If the two spheres are then separated, the sphere that was closer to the glass rod will have a net negative charge and the sphere that was further from the rod will have a net positive charge. This process is called **charging by induction**.

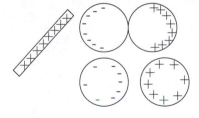

We can think of Earth itself as a large conducting object. When a metal object is brought into contact with the surface of Earth, we say that the object is **grounded**. Earth has an effectively unlimited supply of charges. It can readily absorb or provide as many electrons as are necessary for a given situation.

Insulating objects can also be charged by induction, at least temporarily. If a charged object like a negatively charged rubber balloon is brought near to, but does not touch, an electrically neutral object like a wall, the balloon will attract the protons and repel the electrons in the neutral wall. The neutral object is now **polarized**. Although still electrically neutral, there is an effective separation of the neutral object's positive and negative charges. This effect occurs in both conducting and insulating materials. In tightly bound insulators, the bound electrons are pushed just slightly away from a negatively charged object, but remain bound to their nuclei. In conductors, the effect is much more dramatic, as the free electrons can move as far from the negatively charged object as the geometry of the situation permits. This polarization is generally only temporary. If the negatively charged object is separated a great distance from the neutral object, then the charges of the neutral object will eventually return to a uniformly uncharged distribution.

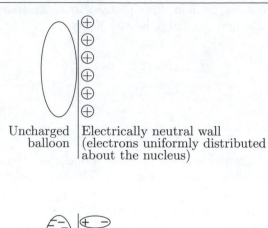

Uncharged balloon | Electrically neutral wall (electrons uniformly distributed about the nucleus)

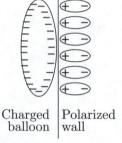

Charged balloon | Polarized wall

Common Pitfalls

> Because of inductive effects, it is possible for a charged object to attract an electrically neutral object.

3. TRUE or FALSE: Because of induction, a charged object can repel an electrically neutral object.

4. After combing your hair with a plastic comb you find that when you bring the comb near a small bit of paper, the bit of paper moves toward the comb. Then, shortly after the paper touches the comb, it moves away from the comb. Explain these observations.

21.3 Coulomb's Law

In a Nutshell

Coulomb's law says that the force exerted by one point charge on another acts along the line between the charges. The force varies inversely as the square of the distance separating the charges and is proportional to the product of the charges. The force is repulsive if the charges have the same sign and attractive if the charges have opposite signs. Mathematically, the force that charge 1 exerts on charge 2 can be written as $\vec{F}_{12} = \dfrac{kq_1q_2}{r_{12}^2}\hat{r}_{12}$, where q_1 and q_2 are the values of the two point charges, k is the Coulomb constant, and $\vec{r}_{12}$ is the vector that points *from* the location of the charge exerting the force (charge 1) *toward* the location of the charge feeling the force (charge 2). $r_{12} = |\vec{r}_{12}|$ is the magnitude (length) of $\vec{r}_{12}$, and $\hat{r}_{12} = \vec{r}_{12}/r_{12}$ is a unit vector in the direction of $\vec{r}_{12}$. The force charge 2 exerts on charge 1 can be written in the same way, but reversing each of the subscripts. $\vec{r}_{21}$ then points *from* the position of charge 2 *toward* the position of charge 1—that is, $\vec{r}_{12} = -\vec{r}_{21}$. The result is that $\vec{F}_{21} = -\vec{F}_{12}$, which we knew had to be the case according to Newton's third law.

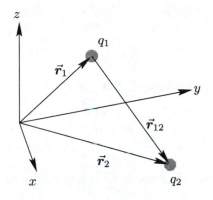

The electric force, like all other forces you have studied, obeys the **principle of superposition**. As a consequence, if several charges exert a force on another charge, the net force on that charge is the vector sum of the individual forces exerted on that charge by all the others.

Physical Quantities and Their Units

Coulomb constant	$k = 8.99 \times 10^9$ N $\cdot$ m^2/C^2

Fundamental Equations

Coulomb's law	$\vec{F}_{12} = \dfrac{kq_1q_2}{r_{12}^2}\hat{r}_{12}$

Common Pitfalls

> ➤ If you are not careful when using Coulomb's law to compute electric force, it is easy to get the signs wrong. You should double-check every result with a quick sanity check: charges of the same sign repel and charges of opposite sign attract. If your answer does not demonstrate this fact, then you need to look for a sign mistake.

5. TRUE or FALSE: As written in Fundamental Equations, Coulomb's law is automatically consistent with Newton's third law.

6. Some days it can be frustrating to attempt to demonstrate electrostatic phenomena for a class. An experiment that works beautifully one day may fail the next day if the weather has changed. Air conditioning helps a lot. Why is this?

Try It Yourself #3

A charge of $+3.00$ μC is located at the origin and a second charge of -2.00 μC is located on the x,y plane at the point (30.0 cm, 20.0 cm). Determine the electric force exerted by the -2.00 μC charge on the 3.00 μC charge.

Picture: Coulomb's force law will be used to calculate the force.

Solve:

Write Coulomb's force law as a calculational guide.	
Draw a sketch of the two charges and include the x and y axes. Label the charge at the origin q_2 and the other charge q_1. Draw the vector $\vec{r}_{12}$, which goes from q_1 to q_2.	
Determine the position vector $\vec{r}_1$ for q_1, using $\hat{\imath}$ and $\hat{\jmath}$ notation.	
Determine the position vector $\vec{r}_2$ for q_2, using $\hat{\imath}$ and $\hat{\jmath}$ notation.	
Calculate the vector $\vec{r}_{12}$.	

Evaluate both r_{12} and $\hat{r}_{12}$.	
Now you have expressions for all values needed to calculate the Coulomb force. Substitute all values, with units, into the expression from the first step to find the vector force.	$\vec{F}_{12} = (0.345 \text{ N})\hat{i} + (0.230 \text{ N})\hat{j}$

Check: Opposite charges should attract, so we would expect the force of q_1 on q_2 to be generally in the $+x$ and $+y$ directions, which it is.

Taking It Further: How would the answer change if both charges were negative?

Try It Yourself #4

Two charges lie on the x axis, a -2.00 μC charge at the origin and a $+3.00$ μC charge at $x = 0.100$ m. At what position x, if any, is the force due to these charges on a $+4.00$ μC charge equal to zero?

Picture: We will use Coulomb's vector force law and the principle of superposition.

Solve:

Draw a sketch of the situation. The $+4.00$ μC charge can, in principle, lie anywhere on the x axis. For simplicity in visualizing, place this charge at some coordinate $x > 0.1$ m.	
Algebraically write Coulomb's force law in vector form to use as a guide.	

Determine algebraic expressions for $\vec{r}$, r, and $\hat{r}$ to use when calculating the force on the $+4.00$ μC charge due to the -2.00 μC charge. Use appropriate subscripts to distinguish these variables from those in the next step.	
Determine algebraic expressions for $\vec{r}$, r, and $\hat{r}$ to use when calculating the force on the $+4.00$ μC charge due to the $+3.00$ μC charge. Use appropriate subscripts to distinguish these variables from those in the previous step.	
Algebraically apply the law of superposition, equating the total force on the 4.00 μC charge to the force from both of the other charges.	
Substitute the appropriate $\vec{r}$, r, and $\hat{r}$ into the above expression. Solve for the coordinate x when the total force is equal to zero.	$x = -0.445$ m

Check: Your solution should have provided two possible values for x. How can you figure out if both or only one is correct? You need to apply your physical reasoning to the problem.

    We know the $+4.00$ μC charge will be attracted to the -2.00 μC charge and repelled by the $+3.00$ μC charge. For $x < 0$, the attractive force is in the $+\hat{\imath}$ direction, and the repulsive force is in the $-\hat{\imath}$ direction. Furthermore, the $+4.00$ μC charge can get closer to the smaller negative charge, so it is possible for the net force to equal zero. Between the two charges, both the attractive and repulsive forces will be in the $+\hat{\imath}$ direction, so it is impossible for the net force to be zero. For $x > 0.1$ m, the attractive force is in the $-\hat{\imath}$ direction, and the repulsive force is in the $+\hat{\imath}$ direction, so you might think that a net force of zero is possible. However, the $+4.00$ μC charge will always be closer to the positive charge, which means that the repulsive force will always be larger than the attractive force, so a net force of zero is not possible.

Taking It Further: Is it possible for a -4.00 μC charge to experience a net force of zero due to these same two charges? Why or why not?

21.4   The Electric Field

In a Nutshell

Action-at-a-distance forces can be more easily understood by introducing the concept of **fields**. In the case of the electric force, charges create **electric fields**, and it is the electric field that actually exerts a force on some other charge. The electric field at any point in space is defined as the force per unit charge on an infinitesimally small test charge. If the test charge is in the presence of only another point charge, we can use this definition of electric field to arrive at Coulomb's law for the electric field. The figure shows **electric field lines** of an isolated positive point charge. The field lines actually start right on the point charge.

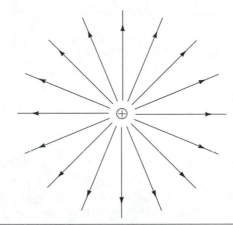

Electric fields, like forces, obey the law of superposition. The electric field at a field point is the vector sum of the electric fields at that point due to all the point charges present.

A system of two point charges that are equal in magnitude, opposite in sign, and separated by a small distance L is called an **electric dipole**. An electric dipole is characterized by its electric dipole moment $\vec{p}$ which is defined as $\vec{p} = q\vec{L}$, where q is the magnitude of one of the charges and $\vec{L}$ is the vector from the location of the negative charge to the location of the positive charge.

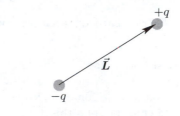

Calculating the Resultant Electric Field

Picture: To calculate the resultant electric field $\vec{E}_P$ at field point P due to a specified distribution of point charges, draw the charge configuration. Include coordinate axes and the field point on the drawing.

Solve:

1. On the drawing label the distance r_{iP} from each charge to point P. Include an electric field vector $\vec{E}_{iP}$ for the electric field at P due to each point charge.
2. If the field point and the point charges are not on a single line, then label the angle each individual electric field vector $\vec{E}_{iP}$ makes with one of the coordinate axes.
3. Calculate the components of each individual field vector $\vec{E}_{iP}$ along the axes directions, and use these to calculate the components of the resultant electric field $\vec{E}_P$.

Physical Quantities and Their Units

Electric field $\vec{E}$ SI units of newtons per coulomb (N/C)

Fundamental Equations

Definition of electric field $\displaystyle \vec{E} = \lim_{q_0 \to 0} \frac{\vec{F}}{q_0}$

Important Derived Results

Force on a test charge in an electric field $\qquad \vec{F} = q_0 \vec{E}$

Electric field of a point charge $\qquad \vec{E}_{iP} = \dfrac{kq_i}{r_{iP}^2} \hat{r}_{iP}$

Superposition of electric fields $\qquad \vec{E}_P = \sum_i \vec{E}_{ip}$

Definition of electric dipole moment $\qquad \vec{p} = q\vec{L}$

Common Pitfalls

> ➤ When the electric fields of two or more point charges superpose, the resultant electric field is the *vector sum* of the individual fields.
> ➤ You have to be forever careful about signs. The force on a positive charge is in the direction of the electric field $\vec{E}$ but the force on a negative charge is directed opposite to $\vec{E}$.
> ➤ An electric dipole consists of two charges that are equal in magnitude and of opposite sign.
> ➤ The equation $\vec{E} = \vec{F}/q_0$ defines the electric field, while the equation $\vec{E}_P = \sum_i (kq_i/r_{iP}^2)\hat{r}_{iP}$ may be used to compute the field due to a distribution of point charges.

7. TRUE or FALSE: Far from a dipole, the magnitude of its electric field decreases with the square of the distance.

8. What are the advantages of thinking of the force on a charge at a point P as being exerted by an electric field at P, rather than by other charges at other locations? Is the convenience of the field as a calculational device worth inventing a new physical quantity? Or is there more to the field concept than this?

Try It Yourself #5

Three point charges are placed on the x,y plane: a $+50.0$-nC charge at the origin, a -60.0-nC charge on the x axis at $x = 10.0$ cm, and a $+150$-nC charge at the point $(10.0 \text{ cm}, 8.00 \text{ cm})$. (a) Find the total electric force on the $+150$-nC charge due to the other two. (b) What is the electric field at the location of the $+150$-nC charge due to the presence of the other two charges?

Picture: For this problem, we will use the superposition of electric forces and fields.

Solve:

The total force on the +150-nC charge is the sum of the forces from the other two charges. Let q_1 be the charge at the origin, q_2 be the -60.0-nC charge, and q_3 be the $+150$-nC charge. Use the superposition principle to write a generic, *algebraic* expression for the total force on q_3.	

Sketch the three charges, making sure to draw $\vec{r}_{13}$ and $\vec{r}_{23}$ properly between the charges.	
Calculate the magnitudes and unit vectors of the $\vec{r}$ vectors of the previous step.	
Substitute all values and their units into the expression found in the first step and solve for the net force on q_3.	$\vec{F}_3 = (3.21 \times 10^{-3}\ \text{N})\hat{\imath} - (1.01 \times 10^{-2}\ \text{N})\hat{\jmath}$
The electric field at the location of the +150-nC charge is the force per unit charge. Use this to find $\vec{E}$.	$\vec{E} = (2.14 \times 10^4\ \text{N/C})\hat{\imath} - (6.72 \times 10^4\ \text{N/C})\hat{\jmath}$

Check: The +150-nC charge will be repelled by the +50.0-nC charge and attracted to the -60.0-nC charge. In the $\hat{\imath}$ direction the repulsion is the only force, so we expect the force on q_3 to have a $+\hat{\imath}$ component. The situation is less clear in the $\hat{\jmath}$ direction because without actually doing the calculations we cannot say for certain whether the attractive or repulsive force will dominate.

Taking It Further: Let the +150-nC charge be replaced with a 432 μC charge at the same location. Given what you already know, what is the quickest way to calculate the force on this new charge due to the other two?

Try It Yourself #6

When a test charge of $+5.00\ \mu C$ is placed at a certain point P, the force that acts on it is 0.0800 N, directed northeast. (a) What is the electric field at P? (b) If the $+5.00\text{-}\mu C$ test charge is replaced by a $-2.00\ \mu C$ charge, what force would act on it?

Picture: For this problem, all you need is the relationship between the electrostatic force and the electric field.

Solve:

Using the vector relationship between force and electric field, find the electric field at point P.	$\vec{E} = 16\,000$ N/C, northeast
Using the electric field from your calculation above, calculate the force on the $-2.00\text{-}\mu C$ charge.	$\vec{F} = 0.0320$ N, southwest

Check: The electric field is the same in both cases, so the force on a negative charge should be in the opposite direction of the force on a positive charge.

Taking It Further: What, if anything, could cause the electric field at point P to change values?

21.5   Electric Field Lines

In a Nutshell

We can draw **electric field lines** to help us visualize the electric field. An individual vector quantity, such as the velocity of a projectile, is best illustrated by a straight arrow pointing in the direction of the vector with a length that is proportional to the magnitude of the vector. However, a vector field, such as the velocity of the water in a flowing stream, is best illustrated by a representative number of directed, curved lines that bend so as to remain tangent to the direction of the vector field. In this representation the number of lines per unit area passing through a small surface oriented at right angles to the field direction is proportional to the local magnitude of the field vector. Arrowheads are drawn on the field lines to show which way the vector points.

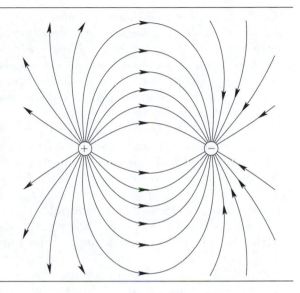

Drawing Field Lines

Picture: Electric field lines emanate from positive charges (or infinity) and terminate on negative charges (or infinity).

Solve:

1. The lines emanating from (or terminating on) an isolated point charge are drawn uniformly spaced as they emanate (or terminate).
2. The number of lines emanating from a positive charge (or terminating on a negative charge) is proportional to the magnitude of the charge.
3. The density of the lines at any point (the number of lines per unit area through a surface element normal to the lines) is proportional to the magnitude of the field there.
4. At large distances from a system of charges that has a nonzero net charge, the field lines are equally spaced and radial, as if they emanated from (or terminated on) a single point charge equal to the total charge of the system.

Check: Make sure that the field lines never intersect. (If two field lines intersected, that would indicate two directions for $\vec{E}$, and thus two directions for the net force—which is impossible, at the point of intersection.)

Common Pitfalls

> When the electric fields of two or more point charges superpose, the resultant electric field is the vector sum of the individual fields.

9. TRUE or FALSE: The electric field due to an electric dipole is always parallel to the direction of the electric dipole moment $\vec{p}$.

10. Sketch electric field lines for the two charges shown below.

+3 pC   −2 pC

21.6 Action of the Electric Field on Charges

In a Nutshell

When a particle with a charge q is placed in an electric field $\vec{E}$ it experiences a force $\vec{F} = q\vec{E}$. If this is the only force acting on the particle, then in accordance with Newton's second law the acceleration of the particle is $\vec{a} = \vec{F}/m = (q/m)\vec{E}$, where m is the mass of the particle.

It is fairly common for electrons to acquire speeds of 10 percent of the speed of light or more. For speeds this high, Newton's laws are inadequate, and the more generally valid laws of the special theory of relativity must be used. For the purposes of this section of the text, however, we will neglect relativistic effects.

Atoms and molecules, even though they are electrically neutral, are affected by electric fields. An atom consists of a small positive nucleus surrounded by one or more electrons. We can think of the atom as a small, massive, positively charged nucleus surrounded by a negatively charged electron cloud. If the electron cloud is spherically symmetric, its center of charge is at the center of the atom, coinciding with the center of the nucleus. Such an atom does not have an electric dipole moment and is said to be **nonpolar**. When a nonpolar atom is placed in an electric field, the force exerted by the field on the negative electron cloud is oppositely directed to the force exerted by the field on the positively charged nucleus. These forces cause the centers of charge of the atom's negative and positive charges to move in opposite directions until the attractive forces the charges exert on each other balance the forces exerted by the external electric field. When this happens the atom is like an electric dipole. The dipole moment of a nonpolar atom or molecule in an external electric field is called an **induced dipole moment**.

In some molecules, water for example, the center of positive charge does not coincide with the center of negative charge even in the absence of an external electric field. These **polar molecules** have a permanent electric dipole moment.

When an electric dipole $\vec{p}$ is placed in a uniform electric field $\vec{E}$ the force on the positive charge and the force on the negative charge are oppositely directed but do not always act along the same line. As shown, these two forces then tend to align the dipole with the electric field direction. The torque vector $\vec{\tau}$ exerted by the electric field on the dipole is given by $\vec{\tau} = \vec{p} \times \vec{E}$.

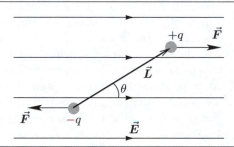

An electric dipole in an external electric field has a minimum amount of potential energy when it is aligned with the electric field. If the potential energy is defined to be zero when $\theta = 90°$, then the potential energy of an electric dipole is given by $U = -\vec{p} \cdot \vec{E}$.

Important Derived Results

Force on a charged particle $\qquad\qquad\qquad \vec{F} = q\vec{E}$

Torque on an electric dipole $\qquad\qquad\qquad \vec{\tau} = \vec{p} \times \vec{E}$

Potential energy of an electric dipole $\qquad\qquad U = -\vec{p} \cdot \vec{E}$

Common Pitfalls

> The electric field points in the direction of the force it exerts on a positive charge, but in the *opposite* direction of the force it exerts on a negative charge.

11. TRUE or FALSE: Because an electric dipole has a zero net charge, a uniform electric field exerts no net force on it.

12. Do electric field lines point along the trajectory of positively charged particles? Why or why not?

Try It Yourself #7

An electron is traveling to the right along the x axis with kinetic energy K, which is along the axis of a cathode ray tube as shown. There is an electric field $\vec{E} = (2.00 \times 10^4 \text{ N/C})\hat{j}$ between the deflection plates, which are 6.00 cm long and are separated by 2.00 cm. Determine the minimum initial kinetic energy the electron can have and still avoid colliding with one of the plates.

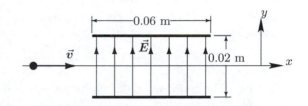

Picture: Determine the electron's acceleration. You will need to use two-dimensional kinematics relationships to find the initial kinetic energy.

Solve:

Find an *algebraic* expression for the acceleration of the electron from its mass and the electrostatic force on it. If the positive y axis is directed upward, then the acceleration of the electron will be in the $-\hat{j}$ direction.	
Write an *algebraic* kinematic expression for the y position of the electron as a function of time. The electron is initially on the x axis, and the initial y component of the velocity is zero.	
The x component of the force is zero, so the electron moves with constant speed in the positive x direction. The initial velocity is entirely in the x direction. Use this information to find an expression for the time t as a function of the electron's x position.	
Algebraically substitute this expression for time into the expression for the electron's y position.	

Rearrange the above expression to solve for the initial kinetic energy of the electron, $K = \frac{1}{2}mv_0^2$.	
Finally, substitute all values with their units into the above expression to find a numerical value for the kinetic energy. The minimum kinetic energy can be arrived at by assuming the electron just hits the right-hand edge of the lower plate. This should give you the appropriate values to use for x and y.	
	$K_{min} = 2.88 \times 10^{-16}$ J

Check: The units work out, but this seems like a pretty low energy. To see if it is reasonable, see Taking It Further, below. This example demonstrates that for real, fast-moving electrons, considerable electric fields, on the order of 1000× larger than in this problem, are required to divert electrons over short distances.

Taking It Further: For kinetic energies this small, you might expect a pretty low speed. Calculate the initial speed of the electron and explain the result.

Try It Yourself #8

The electric field is zero everywhere except in the region $0 \leq x \leq 3.00$ cm, where there is a uniform electric field of 100 N/C in the $+y$ direction. A proton is moving along the negative x axis with a velocity $\vec{v} = (1.00 \times 10^6$ m/s$)\hat{\imath}$. When the proton passes through the region $0 \leq x \leq 3.00$ cm the electric field exerts a force on it. (a) When the x coordinate of the proton's position is 3.00 cm, what is its velocity and what is the y coordinate of its position? (b) When the x coordinate of its position equals 10.0 cm, what is its velocity and what is the y coordinate of its position?

Picture: This is a two-segment problem. For the first segment, find the acceleration of the proton while it is in the region of the electric field. Use two-dimensional kinematics to find the velocity of the proton just at the point it leaves the region with an electric field. Once the proton leaves the electric field, there will be no force acting on it, so it will not experience any acceleration, just constant-velocity two-dimensional motion.

Solve:

Draw a sketch to help you visualize the problem.	

Find an *algebraic* expression for the acceleration of the proton while under the influence of the electric field.	
From the equations for motion in the x direction, find an *algebraic* expression for the time it will take the proton to travel from $x = 0$ cm to $x = 3$ cm.	
From the equations for motion in the y direction, determine the y component of the velocity when the electron is at $x = 3$ cm. The x component of the velocity is the same as before. Why?	$\vec{v}(x = 3 \text{ cm}) = (1.00 \times 10^6 \text{ m/s})\hat{\imath} + (288 \text{ m/s})\hat{\jmath}$
From the equations for motion in the y direction, determine the y position of the electron when $x = 3$ cm.	$y = 4.32 \times 10^{-6}$ m
Determine the acceleration of the proton after it leaves the electric field region, and use this to calculate the velocity of the electron when it is at $x = 10$ cm.	$\vec{v}(x = 10 \text{ cm}) = (1.00 \times 10^6 \text{ m/s})\hat{\imath} + (288 \text{ m/s})\hat{\jmath}$
From the equations for motion in the x direction determine the time required for the electron to move from $x = 3$ cm to $x = 10$ cm and use this time to find the y position of the electron when it is at $x = 10$ cm.	$y(x = 10 \text{ cm}) = 2.45 \times 10^{-5}$ m

Check: The units all work out.

Taking It Further: If the proton is replaced with an electron, how would your answers to this problem change qualitatively?

QUIZ

1. TRUE or FALSE: Positively charged particles repel each other, whereas negatively charged particles attract each other.

2. TRUE or FALSE: If an otherwise free point charge is in a region in which there is an electric field $\vec{E}$, its acceleration is necessarily along the direction of the field.

3. After combing your hair with a plastic comb you find that when you bring the comb near an empty aluminum soft-drink can that is lying on its side on a nonconducting tabletop, the can rolls toward the comb. After being touched by the comb the can is still attracted by the comb. Explain.

4. At point P in the figure, $\vec{E}$ is found to be zero. What can you say about the signs and magnitudes of the charges?

5. A positively charged glass rod attracts a lighter object suspended by a thread. Does it follow that the object is negatively charged? If, instead, the rod repels it, does it follow that the suspended object is positively charged?

6. An electron, released in a region where the electric field is uniform, is observed to have an acceleration of 3.00×10^{14} m/s^2 in the positive x direction. Determine the electric field producing this acceleration. Assuming the electron is released from rest, determine the time required for it to reach a speed of 11,200 m/s, the escape speed from Earth's surface.

7. A charge $q_1 = +2q$ is at the origin and a charge $q_2 = -q$ is on the x axis at $x = a$. Find expressions for the total electric field on the x axis in each of the regions (a) $x < 0$; (b) $0 < x < a$; and (c) $a < x$. (d) Determine all points on the x axis where the electric field is zero. (e) Make a plot of E_x versus x for all points on the x axis, $-\infty < x < \infty$.

Chapter 22

The Electric Field II: Continuous Charge Distributions

22.1 Calculating $\vec{E}$ from Coulomb's Law

In a Nutshell

We understand that charge is quantized on a microscopic scale. However, situations often occur in which it is helpful to treat charges as if their spatial distribution were continuous. This approximation is analogous to the manner in which mass is often considered to be continuously distributed, even though it is accepted knowledge that most of the mass of an atom is located in the atomic nucleus, which is extremely small compared to the atom as a whole.

When we calculate the electric field at some point P due to continuous charge distributions, we must divide that distribution into small differential pieces, each of which has a small differential amount of charge which can be treated as a point source. Charges are either distributed along a line, on a surface, or throughout a volume. We can write the differential amount of charge for each case as $dq = \lambda\, dL$, $dq = \sigma\, dA$, or $dq = \rho\, dV$, respectively. Here, dL, dA, and dV are differential length, area, and volume elements, and λ, σ, and ρ are the charge per unit length, area, and volume, respectively.

Once a differential charge element has been identified, its electric field at point P is calculated in the same manner we calculated the electric field from discrete point charges. The only difference is that we are now explicitly calculating a small differential amount of electric field due to the small differential charge distribution: $d\vec{E} = \dfrac{k\, dq}{r^2}\hat{r}$. To calculate the electric field due to an entire charge distribution, we integrate this expression over the entire charge distribution: $\vec{E} = \displaystyle\int d\vec{E} = \int \dfrac{k\hat{r}}{r^2}\, dq$. In these expressions, $\vec{r}$, $\hat{r}$, and r have the same meaning as in Chapter 21, applied to the differential charge element.

The Coulomb constant k is often written in terms of another constant, the **permittivity of free space** ϵ_0: $k = 1/(4\pi\epsilon_0)$.

Calculating $\vec{E}$ via Integration

Picture: Sketch the charge configuration along with a field point P (the point where $\vec{E}$ is to be calculated). The sketch should include an increment of charge dq at an arbitrary source point S.

Solve:

1. Add coordinate axes to the sketch. The choice of axes should exploit any symmetry of the charge configuration. For example, if the charge is along a straight line, select that line as one of the coordinate axes. Draw a second axis that passes through the field point P. In addition, include the coordinates of both P and S, the distance r between P and S, and the unit vector $\hat{r}$ directed away from S toward P.

2. To compute the electric field $\vec{E}$ by integration, we express $d\vec{E} = dE_r \hat{r}$ in component form. The x component of $d\vec{E}$ is $dE_x = dE_r \hat{r} \cdot \hat{i} = dE_r \cos\theta$, where θ is the angle between $\hat{r}$ and $\hat{i}$, and the y component of $d\vec{E}$ is $dE_y = dE_r \hat{r} \cdot \hat{j} = dE_r \sin\theta$.

3. Express $\vec{E}$ in terms of its x and y components:

$$E_x = \int dE_x = \int dE_r \cos\theta = \int \frac{k\,dq}{r^2} \cos\theta$$

$$E_y = \int dE_y = \int dE_r \sin\theta = \int \frac{k\,dq}{r^2} \sin\theta$$

4. To calculate E_x, express dq as $\rho\,dV$, or $\sigma\,dA$, or $\lambda\,dL$ (whichever is appropriate) and integrate. To calculate E_y, follow a procedure similar to that used for calculating E_x.

5. Symmetry arguments are sometimes used to show that one or more components of $\vec{E}$ are equal to zero.

Check: If the charge distribution is confined to a finite region of space and the net charge is non-zero, then at points far from the charge distribution the expression for the electric field should approach that of a point charge located at the center of charge. (If the charge configuration is sufficiently symmetric then the location of the center of charge can be obtained by inspection.)

Physical Quantities and Their Units

Coulomb's constant

$$k = 8.99 \times 10^9 \text{ N} \cdot \text{m}^2/\text{C}^2$$

Permittivity of free space

$$\epsilon_0 = \frac{1}{4\pi k} = 8.85 \times 10^{-12} \text{ C}^2/(\text{N} \cdot \text{m}^2)$$

Charge per unit volume

$$\rho = \frac{dq}{dV} \text{ with units of C/m}^3$$

Charge per unit area

$$\sigma = \frac{dq}{dA} \text{ with units of C/m}^2$$

Charge per unit length

$$\lambda = \frac{dq}{dL} \text{ with units of C/m}$$

Fundamental Equations

Differential electric field

$$d\vec{E} = \frac{k\,dq}{r^2} \hat{r}$$

Electric field due to a continuous charge distribution
$$\vec{E} = \int d\vec{E} = \int \frac{k\hat{r}}{r^2} dq$$

Important Derived Results

Electric field on axis of a uniformly charged ring $\quad E_z = \dfrac{kQz}{(z^2 + a^2)^{3/2}}$

Electric field of a large, uniform plane of charge $\quad E_z = \text{sign}(z) \cdot \dfrac{\sigma}{2\epsilon_0}$

Common Pitfalls

➤ When calculating the electric field via integration you need to evaluate it one component at a time.

➤ In electrostatics the formulas for the perimeter, surface area, and the volume of common figures such as the circular disk, the sphere, and the cylinder are frequently encountered. If you need to review these formulas, you can find them in Appendix B on page AP-5 of the text.

➤ The charge density is given by the total charge divided by the total length, area, or volume, whichever is appropriate, *only* if the charge is uniformly distributed.

➤ If the charge is not uniformly distributed then you need to use the provided expression that describes the distribution of charge.

➤ Remember that differential quantities also have units. In the SI system that means, for example, that dx has units of meters, and $dA = dx\,dy$ has units of m^2.

1. TRUE or FALSE: Near an isolated uniformly charged plane, the magnitude of the electric field decreases with the first power of the distance from the plane.

2. Explain why integrating over a charge distribution works to find the net electric field at some point P due to that charge distribution.

Try It Yourself #1

Find the electric field at point P on the y axis at $y = 3.00$ m due to the charge on a wire on the x axis between $x = 1.00$ m and $x = 3.00$ m whose uniform charge per unit length is 4.00 μC/m.

Picture: Sketch the situation to help with the visualization. You will need to integrate the electric field produced at P due to small, differential charge elements along the charged wire. Because the wire lies along the x axis, your integration should end up being with respect to dx.

Solve:

Sketch the line of charge along with the field point P. Draw an increment of charge dq at an arbitrary source point S. Add coordinate axes to the sketch. Include the coordinates of both P and S (the location of your differential charge element), the distance r between P and S, and the unit vector $\hat{r}$ directed away from S toward P.	

Write the integral form of Coulomb's law in *algebraic* vector component form as a guide in solving the rest of the problem.	
Determine an expression for the charge element dq. As you move from one differential charge element to the next, you move along the x axis, so your expression should include a dx.	
From your sketch determine an expression for $\vec{r}$ in component form for the dq that you drew. Remember this vector points *from* the source of the electric field *toward* the field point. Because $\vec{r}$ is different for each differential charge element, at least one component of your vector must contain the integration variable, in this case x.	
Determine the magnitude of $\vec{r}$ found in the previous step.	
Determine $\hat{r}$ from the previous two steps.	
Substitute all the expressions for each variable into Coulomb's law. Note that because $\hat{r}$ has both an $\hat{i}$ and a $\hat{j}$ component, the electric field will also potentially have both components.	

Integrate the $\hat{\imath}$ term to find the x component of the electric field at P. Your limits of integration are determined by the location of the charged wire.	
	$E_x = -2900 \text{ N/C}$
Integrate the $\hat{\jmath}$ term to find the y component of the electric field at P. Your limits of integration are determined by the location of the charged wire.	
	$E_y = +4690 \text{ N/C}$

Check: Because point P is "up" and "to the left" of the positive charge distribution, we expect the electric field to also point in that direction, which it does.

Taking It Further: Compare the results of this problem to the situation of having all the charge on the wire located at the center of the wire.

Try It Yourself #2

A semicircular ring of charge with radius $r = 3$ cm and centered on the origin is situated as shown. The charge is not uniform, but is given by $\lambda = -(3\ \mu\text{C/m})\cos\theta$, where θ is measured from the $+y$ axis as indicated. Find the electric field due to this charge distribution at point P located at $y = -4$ cm on the y axis.

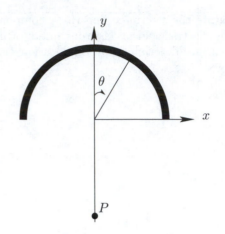

Picture: This is a linear charge density because the charge is along a line. It does not matter that the line is curved. We will need to integrate Coulomb's law to find the electric field at P.

Solve:

Write *algebraically* a generic expression for Coulomb's law for continuous charge distributions. We will use this to guide us through the calculation.	
On the figure depicting the situation, draw an increment of charge dq at an arbitrary source point S. Draw $\vec{r}$, which is from S to P.	
Determine an *algebraic* expression for the charge element dq. As you move from one differential charge element to the next, you move around the semicircular arc, which is done by changing θ, so your expression should include a $d\theta$. Your expression must have units of coulombs. Careful dimensional analysis should help you determine what other variables are needed.	
From your sketch determine an expression for $\vec{r}$ in component form for the dq that you drew. Remember this vector points *from* the source of the electric field *toward* the field point. Because $\vec{r}$ is different for each differential charge element, at least one component of your vector must contain the integration variable, in this case θ. Be careful here, because θ is not defined in the usual way.	
Determine the magnitude of $\vec{r}$ found in the previous step.	

Determine $\hat{r}$ from the previous two steps.	
Substitute all the expressions for each variable into Coulomb's law. Note that because $\hat{r}$ has both an $\hat{i}$ and a $\hat{j}$ component, the electric field will also potentially have both components.	
Integrate the $\hat{i}$ term to find the x component of the electric field at P. Your limits of integration are determined by the location of the charged wire.	$E_x = 0 \text{ N/C}$
Integrate the $\hat{j}$ term to find the y component of the electric field at P. Your limits of integration are determined by the location of the charged wire. You may need more paper to evaluate this integral.	$E_y = +4.57 \times 10^6 \text{ N/C}$

Check: The charge distribution is negative, so the electric field must point *toward* the charges. Our result has a $+y$ component which does just this. Even though the charge is not uniformly distributed it is, however, symmetric about the y axis because $\cos\theta$ is an even function. As a result, you could have expected the x component of the electric field to be zero.

Taking It Further: If the semicircle were extended to be a complete ring, with the same function for the linear charge distribution, describe qualitatively how you would expect the electric field at point P to change, if at all.

22.2   Gauss's Law

In a Nutshell

The **electric flux** ϕ is a mathematical quantity that is proportional to the number of electric field lines that penetrate a surface. The larger the flux, the more field lines that penetrate the surface. The electric flux is defined to be $\phi = \int_S d\phi = \int_S \vec{E} \cdot \hat{n}\, dA$. In this expression $\hat{n}$ is a unit vector perpendicular (normal) to the differential area element dA.

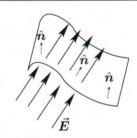

The **electric flux through closed surface** is defined to be $\phi_{\text{net}} = \oint_S \vec{E} \cdot \hat{n}\, dA$. By convention, the unit vector $\hat{n}$ points out of the surface at each point on the surface. As a result, the local flux is positive where field lines point out of the surface and negative where the field lines point into the surface.

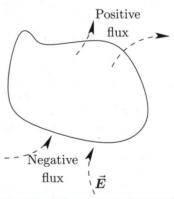

Gauss's Law states that the electric flux through a closed surface is equal to the net charge enclosed by that surface divided by ϵ_0. Gauss's law and Coulomb's law give the same results for the electric field in electrostatics. However, Gauss's law is more general because it is also valid when determining electric fields from moving charge distributions. Coulomb's law is not valid in that case.

Physical Quantities and Their Units

Electric flux   SI units of $\text{N} \cdot \text{m}^2/\text{C}$

Fundamental Equations

Gauss's law $\oint_S \vec{E} \cdot \hat{n}\, dA = \dfrac{Q_{\text{inside}}}{\epsilon_0}$

Common Pitfalls

> ➤ If a closed surface encloses zero net charge, that does *not* mean the electric field is zero everywhere on the surface. In fact, there is typically a nonzero electric field present everywhere on the surface. What *is* zero is the sum of the electric fluxes over the entire surface.

3. TRUE or FALSE: Electric flux is a vector and has a direction associated with it.

4. Is the electric field $\vec{E}$ in Gauss's law only that part of the electric field due to the charge inside a surface, or is it the total electric field due to all charges both inside and outside the surface? Explain.

Try It Yourself #3

The figure shows a prism-shaped surface that is 40.0 cm high (parallel to the y axis), 30.0 cm deep (parallel to the x axis), and 80.0 cm long (parallel to the z axis). The prism is immersed in a uniform electric field of $(500\ \text{N/C})\hat{\imath}$. Calculate the electric flux out of each of its five faces and the net electric flux out of the entire closed surface.

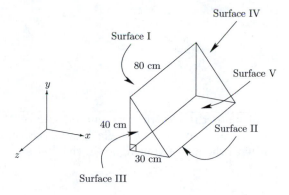

Picture: There is no net flux out of any face where $\vec{E} \cdot \hat{n} = 0$. The total flux equals the sum of the fluxes out of each face.

Solve:

Draw a sketch of each of the faces of the prism, including the direction of $\hat{n}$ for each face and $\vec{E}$ on each face.	
Calculate the electric flux out of surface I.	$\phi_{\text{I}} = -160\ \text{N} \cdot \text{m}^2/\text{C}$

Calculate the electric flux out of surface II.	
	$\phi_{\text{II}} = 0 \text{ N} \cdot \text{m}^2/\text{C}$
Calculate the electric flux out of surface III.	
	$\phi_{\text{III}} = 0 \text{ N} \cdot \text{m}^2/\text{C}$
Calculate the electric flux out of surface IV.	
	$\phi_{\text{IV}} = 0 \text{ N} \cdot \text{m}^2/\text{C}$
Calculate the electric flux out of surface V. Find the cosine of the angle between $\vec{E}$ and $\hat{n}$ with the aid of your figure in the first step.	
	$\phi_{\text{V}} = +160 \text{ N} \cdot \text{m}^2/\text{C}$
Calculate the net electric flux out of the prism, which is the sum of the fluxes.	
	$\phi_{\text{net}} = 0 \text{ N} \cdot \text{m}^2/\text{C}$

Check: Since there is no charge inside the prism, we expect the net flux through the walls of the prism to be equal to zero.

Taking It Further: If in addition to the given electric field the prism also enclosed a point charge of $-2.00 \ \mu\text{C}$, qualitatively how would your answers above change, if at all?

Try It Yourself #4

The prism of the previous problem is now immersed in a uniform electric field of $\vec{E} = (500\text{ N/C})\hat{\imath} + (400\text{ N/C})\hat{\jmath}$. Calculate the electric flux out of each of its five faces and the net electric flux out of the entire closed surface.

Picture: Use the sketch of each face of the prism from the previous example, and calculate the flux through each face, as before.

Solve:

Calculate the electric flux out of surface I.	$\phi_{\text{I}} = -160\text{ N} \cdot \text{m}^2/\text{C}$
Calculate the electric flux out of surface II.	$\phi_{\text{II}} = -96.0\text{ N} \cdot \text{m}^2/\text{C}$
Calculate the electric flux out of surface III.	$\phi_{\text{III}} = 0\text{ N} \cdot \text{m}^2/\text{C}$
Calculate the electric flux out of surface IV.	$\phi_{\text{IV}} = 0\text{ N} \cdot \text{m}^2/\text{C}$
Calculate the electric flux out of surface V. Find the cosine of the angle between $\vec{E}$ and $\hat{n}$ with the aid of your figure in the first step.	$\phi_{\text{V}} = +256\text{ N} \cdot \text{m}^2/\text{C}$

Calculate the net electric flux out of the prism, which is the sum of the fluxes.	
	$\phi_{\text{net}} = 0 \text{ N} \cdot \text{m}^2/\text{C}$

Check: Again, since there is no charge inside the prism, we expect the net flux through the walls of the prism to be equal to zero.

Taking It Further: How would your answer change if the Gaussian surface were a sphere rather than the prism?

22.3   Using Symmetry to Calculate $\vec{E}$ with Gauss's Law

In a Nutshell

Gauss's law is always valid. However, it useful *only* when calculating the electric field of symmetrical charge distributions. A charge configuration has **cylindrical** (or **line**) **symmetry** if the charge density depends only on the distance from a line, **planar symmetry** if the charge density depends only on the distance from a plane, and **spherical** (or **point**) **symmetry** if the charge density depends only on the distance from a point.

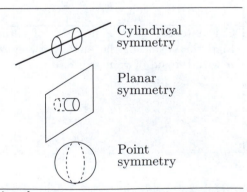

Cylindrical symmetry

Planar symmetry

Point symmetry

Gauss's law can be used to derive Coulomb's law for point charges.

Calculating $\vec{E}$ using Gauss's Law

Picture: Determine whether or not the charge configuration belongs to one of the three symmetry classes. If it does not, then try another method to calculate the electric field. If it does, then sketch the charge configuration and establish the magnitude and direction of the electric field $\vec{E}$ using symmetry considerations.

Solve:

1. On the sketch draw an imaginary closed surface, called a **Gaussian surface**. This surface is chosen so that on each piece of the surface $\vec{E}$ is either zero, normal to the surface with E_n the same everywhere on the piece, or parallel to the surface ($E_n = 0$) everywhere on the piece. For a configuration that has cylindrical (line) symmetry, the Gaussian surface is a cylinder coaxial with the symmetry line. For a configuration that has planar symmetry, you can use a cylinder bisected by the symmetry plane and with its symmetry axis normal to the symmetry plane as a Gaussian surface. For a configuration that has spherical (point) symmetry, the Gaussian surface is a sphere centered on the symmetry point. On each piece of the Gaussian surface sketch an area element dA, an outward normal $\hat{n}$, and the electric field $\vec{E}$.

2. Closed cylindrical surfaces are piecewise continuous, with the surface divided into three pieces. Spherical surfaces consist of a single piece. The flux through each piece of a properly chosen Gaussian surface equals $E_n A$, where E_n is the component of $\vec{E}$ normal to the piece and A is the area of the piece. Add the fluxes through each piece to obtain the total outward flux through the closed surface.

3. Calculate the total charge inside the Gaussian surface.

4. Apply Gauss's law to relate E_n to the charges inside the closed surface and solve for E_n.

Fundamental Equations

Gauss's law
$$\oint_S \vec{E} \cdot \hat{n}\, dA = \frac{Q_{\text{inside}}}{\epsilon_0}$$

Important Derived Results

Electric field of a thin spherical shell
$$E_r = \begin{cases} \dfrac{Q}{4\pi\epsilon_0 r^2} & r > R \\ 0 & r < R \end{cases}$$

Electric field of a uniformly charged sphere
$$E_r = \begin{cases} \dfrac{Q}{4\pi\epsilon_0 r^2} & r \geq R \\ \dfrac{Qr}{4\pi\epsilon_0 R^3} & r \leq R \end{cases}$$

Common Pitfalls

> When calculating the electric flux out of a Gaussian cylinder, remember that you *must always* *explicitly* include the flux through three surfaces: the side of the cylinder and each end of the cylinder. The flux is likely to be zero through at least one surface and nonzero through the other surface(s).

5. TRUE or FALSE: Inside an isolated, infinitely long, uniformly charged cylindrical shell, the magnitude of the electric field is everywhere zero.

6. Inside a spherical charge distribution of constant volume charge density, why is it that as one moves out from the center, the electric field increases as r rather than decreases as $1/r^2$?

Try It Yourself #5

A very long, hollow, charged cylinder has an inner radius of $a = 3.00$ cm, an outer radius of $b = 5.00$ cm, and a uniform charge density of $\rho = +42.0\ \mu C/m^3$. Find the electric field for all radii $r \leq a$, $a \leq r \leq b$ and $b \leq r$.

Picture: Since we have a charged cylinder, we do have cylindrical symmetry, so Gauss's law is appropriate. We will need three separate Gaussian surfaces because we need to calculate the electric field in three distinct regions of space.

Solve:

Sketch the charge distribution. Draw a finite-sized imaginary Gaussian surface at some radius $r_G < a$. Draw the normal vector for each piece of this closed surface. Sketch a predicted direction of the electric field on each piece of that surface.	
Write Gauss's law as a guide for your calculations.	
Using symmetry arguments—state those arguments explicitly here—find an *algebraic* expression for the electric flux through each end of your Gaussian cylinder in terms of the radial component of the electric field E_r.	
Using symmetry arguments—state those arguments explicitly here—find an *algebraic* expression for the electric flux through the curved side of your Gaussian cylinder in terms of the radial component of the electric field E_r.	
Combine the previous two expressions to find an *algebraic* expression for the total flux through your Gaussian cylinder.	

Find an *algebraic* expression for the total charge enclosed by your Gaussian cylinder of radius $r_G < a$, using the relationship between charge density and volume.	
You now have everything you need to solve for E_r as a function of r_G inside the hollow of the cylinder. Substitute the results of your previous two steps into Gauss's law, insert the given values and their units, and solve.	$E_r = 0$
Draw on your sketch a new finite-sized imaginary Gaussian surface at some radius $a < r_G < b$. Draw the normal vector for each piece of this closed surface. Sketch a predicted direction of the electric field on each piece of that surface.	
Using symmetry arguments—state those arguments explicitly here—find an *algebraic* expression for the electric flux through each end of this second Gaussian cylinder in terms of the radial component of the electric field E_r.	
Using symmetry arguments—state those arguments explicitly here—find an *algebraic* expression for the electric flux through the curved side of this second Gaussian cylinder in terms of the radial component of the electric field E_r.	
Combine the previous two expressions to find an *algebraic* expression for the total flux through this second Gaussian cylinder.	
Find an *algebraic* expression for the total charge enclosed by this second Gaussian cylinder of radius $a < r_G < b$, using the relationship between charge density and volume. Don't forget the cylinder is hollow.	

You now have everything you need to solve for E_r as a function of r_G in the shell itself. Substitute the results of your previous two steps into Gauss's law, insert the given values and their units, and solve.	$$E_r = \frac{(2.37 \times 10^6 \ \text{N}/(\text{C} \cdot \text{m}))(r_G^2 - a^2)}{r_G}$$
Draw on your sketch a third finite-sized imaginary Gaussian surface at some radius $b < r_G$. Draw the normal vector for each piece of this closed surface. Sketch a predicted direction of the electric field on each piece of that surface.	
Using symmetry arguments—state those arguments explicitly here—find an *algebraic* expression for the electric flux through each end of this third Gaussian cylinder in terms of the radial component of the electric field E_r.	
Using symmetry arguments—state those arguments explicitly here—find an *algebraic* expression for the electric flux through the curved side of this third Gaussian cylinder in terms of the radial component of the electric field E_r.	
Combine the previous two expressions to find an *algebraic* expression for the total flux through this third Gaussian cylinder.	
Find an *algebraic* expression for the total charge enclosed by this third Gaussian cylinder of radius $b < r_G$, using the relationship between charge density and volume. Don't forget the cylinder is hollow.	

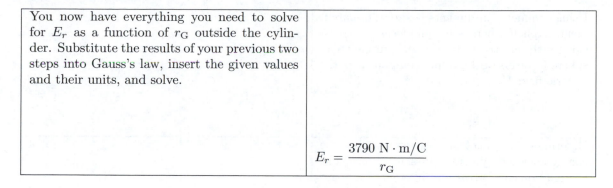

You now have everything you need to solve for E_r as a function of r_G outside the cylinder. Substitute the results of your previous two steps into Gauss's law, insert the given values and their units, and solve.	$E_r = \dfrac{3790 \text{ N} \cdot \text{m/C}}{r_G}$

Check: There should be no electric field in the hollow portion of the long cylinder since there is no charge present there.

Taking It Further: Did you notice anything in common about the actual mathematics involved in calculating the electric field in each region of space? Does this commonality make sense? Explain.

Try It Yourself #6

A sphere of radius R carries a charge distribution that varies with radius according to $\rho = (-23.0 \ \mu\text{C/m}^4)r$. Find an expression for the electric field at all radii from the center of the sphere.

Picture: Since we have a charged sphere whose charge distribution varies only with r, we still have spherical symmetry and Gauss's law is appropriate. We will need two Gaussian surfaces because we have two regions of space: inside and outside the sphere.

Solve:

Sketch the charge distribution. Draw a finite-sized imaginary Gaussian surface at some radius $r_G < R$. Draw a few normal vectors for this closed surface. Sketch a predicted direction of the electric field on this surface.	
Write Gauss's law as a guide for your calculations	

Using symmetry arguments—state those arguments explicitly here—find an *algebraic* expression for the electric flux through your Gaussian sphere in terms of the radial component of the electric field E_r.	
Determine an *algebraic* expression for the charge enclosed by your Gaussian surface. Use the relationship between a differential amount of charge, the charge density, and the differential volume. Actually carry out the integral here.	
You now have everything you need to solve for E_r as a function of r_G inside the sphere. Substitute the results of your previous two steps, complete with units, into Gauss's law and solve.	$E_r = \left(-6.49 \times 10^5 \ \text{N}/(\text{m}^2 \cdot \text{C})\right) r_G^2$
Draw a second finite-sized imaginary Gaussian surface at some radius $r_G > R$. Draw a few normal vectors for this closed surface. Sketch a predicted direction of the electric field on this surface.	
Using symmetry arguments—state those arguments explicitly here—find an *algebraic* expression for the electric flux through this second Gaussian sphere in terms of the radial component of the electric field E_r.	
Determine an *algebraic* expression for the charge enclosed by the second Gaussian surface. Use the relationship between a differential amount of charge, the charge density, and the differential volume. Actually carry out the integral here.	

You now have everything you need to solve for E_r as a function of r_G outside the sphere. Substitute the results of your previous two steps, complete with units, into Gauss's law and solve.	
	$$E_r = \frac{\left(-6.49 \times 10^5 \text{ N/(m}^2 \cdot \text{C)}\right) R^4}{r_G^2}, \quad r_G \geq R$$

Check: Since electric field is a physical quantity, it should be continuous. Do you get the same value for the radial component of the electric field at $r_G = R$ for both expressions? You should.

Taking It Further: Assume all the charge of the sphere is located at the center of the sphere. Determine the electric field of the resulting point charge using Coulomb's law and compare it to the electric field outside the sphere that you calculated in this problem. Explain your result.

22.4 Discontinuity of E_n

In a Nutshell

The normal component of the electric field changes by σ/ϵ_0 across any surface with a finite surface charge density σ. This can be derived from Gauss's law.

Important Derived Results

Discontinuity of E_n at a surface charge $\qquad\qquad E_{n+} - E_{n-} = \dfrac{\sigma}{\epsilon_0}$

Common Pitfalls

> In real physical situations the electric field is always continuous because we can never create an infinitesimally thin surface layer of charge. However, for mathematical simplicity we consider here only these perfect surfaces of charge. In reality, as long as we are a reasonable distance from the charge layer (maybe a few hundred micrometers), this approximation is actually quite good.

> If there is a region of space with a finite volume charge density, like a charged insulating sphere, then there is no discontinuity in the electric field at the edge of the distribution.

22.5   Charge and Field at Conductor Surfaces

In a Nutshell

For a conductor in electrostatic equilibrium the electric field within the conducting material itself is zero.
For a conductor in electrostatic equilibrium Gauss's law can be used to show that any net charge must reside on the surfaces of the conductor.

Important Derived Results

E_n just outside the surface of a conductor $\qquad\qquad E_n = \dfrac{\sigma}{\epsilon_0}$

Common Pitfalls

> Charge layers can exist on *any* surface of a conductor, both inner and outer surfaces.

7. TRUE or FALSE: The tangential component of the electric field is zero at all points just outside the surface of a conductor in electrostatic equilibrium.

8. The electric field just outside the surface of a conductor is twice that due to an infinite uniformly charged plane having the same surface charge density. Why aren't they the same? Why the factor of 2?

Try It Yourself #7

At all points just outside the surface of a 2.00-cm-diameter steel ball bearing, there exists an electric field of magnitude 400 N/C. Assuming the ball bearing is in electrostatic equilibrium, (a) what is the total charge on the ball? (b) What is the surface charge density on the ball?

Picture: Use Gauss's law to find the total charge of the ball. Then use what we know about conductors to find the surface charge density.

Solve:

Draw a sketch of the ball bearing. Also draw a spherical Gaussian surface just outside the ball bearing, at a radius of r_G. Draw a few normal vectors for this closed surface. Sketch a predicted direction of the electric field on this surface.	

Write Gauss's law as a guide for your calculations	
Using symmetry arguments—state those arguments explicitly here—find an *algebraic* expression for the electric flux through your Gaussian sphere in terms of the radial component of the electric field E_r.	
Equate this flux to the enclosed charge divided by ϵ_0, and substitute values and their units to solve for the enclosed charge, which must be the charge on the ball bearing.	$Q_{\text{ball bearing}} = 1.78 \times 10^{-11}$ C
We know that the electric field inside a conductor is zero, and that all the charge sits on the surface. Since the bearing is solid, the only surface is the outside surface. Use the relationship between the discontinuity of electric field and the surface charge density to find σ.	$\sigma = 3.54 \times 10^{-9}$ C/m^2

Check: Calculate the charge density by dividing the total charge by the surface area of the ball bearing. You better get the same answer.

Taking It Further: How would this problem change if instead of a solid ball bearing we had a thin hollow spherical shell? Explain.

Try It Yourself #8

A -3.20 μC charge sits in the center of a conducting spherical shell in static equilibrium with inner radius 2.50 cm and outer radius 3.50 cm. The shell has a net charge of -5.80 μC. Determine the charge on each surface of the shell and the electric field just outside the shell.

Picture: Draw a sketch of the situation. We will use Gauss's law and the fact that the electric field inside a conducting material in static equilibrium is zero to solve the problem.

Solve:

Draw a sketch showing the charge and the inner and outer radii of the shell.	
Sketch a spherical Gaussian surface with a radius just larger than the inner radius of the shell. Since the surface is in the conducting material itself, determine the electric field on the Gaussian surface, and hence the electric flux through that surface.	
Use Gauss's law to determine the total charge that must be enclosed by that Gaussian surface. We know that there is $-3.20\ \mu$C at the center, so there must be some additional charge also enclosed. This charge sits on the inner surface of the sphere.	$Q_{\text{inner}} = +3.20\ \mu$C
We know that all excess charge lies on the surfaces of the conductor. We know that the inner surface has a total charge of $+3.20\ \mu$C. So what charge must sit on the outer surface to achieve the total charge of the shell?	$Q_{\text{outer}} = -9.00\ \mu$C
Draw a second Gaussian surface with a radius just outside the shell. Draw some normal vectors on this surface, as well as some electric-field vectors.	
Using symmetry arguments—state those arguments explicitly here—find an *algebraic* expression for the electric flux through this second Gaussian sphere in terms of the radial component of the electric field E_r.	

Equate this flux to the total enclosed charge divided by ϵ_0, and rearrange to solve for E_r just outside the sphere. Substitute values, complete with units, into Gauss's law and solve.	
	$E_r = -6.6 \times 10^7$ N/C

Check: You can also calculate the charge on the inner surface by finding the field from just the center charge at a radius just smaller than the inner radius and using the discontinuity of electric fields to find the surface charge density, from which you can find the total charge. You should get the same answer.

Taking It Further: How would this problem change, if at all, if we substituted an insulating shell for a conducting shell?

22.6   The Equivalence of Gauss's Law and Coulomb's Law in Electrostatics*

In a Nutshell

Gauss's law can be derived mathematically from Coulomb's law for the electrostatic case. See the text for more details.

---

*Optional Material

QUIZ

1. TRUE or FALSE: If the net electric flux out of a closed surface is zero, the electric field must be zero everywhere on the surface.

2. TRUE or FALSE: The normal component of the electric field is the same at all points just outside the surface of a conductor in electrostatic equilibrium.

3. If the net electric flux out of a closed surface is zero, does that mean the charge density must be zero everywhere inside the surface? Explain.

4. Explain why the normal component of the electric field just outside a conductor in static equilibrium is equal to σ/ϵ_0.

5. Why is the expression $E = 2\pi k\sigma \left[1 - |x|/\sqrt{x^2 + R^2}\,\right]$ for the electric field along the axis of a uniformly charged thin disk different from the expression $E = 2\pi k\sigma$ for the electric field of a uniformly charged infinite plane when Gauss's law *seemingly* gives the same result for these two cases?

6. A uniform line of charge with linear charge density $\lambda = 4.00$ nC/m extends from $x = -2.00$ m to $x = +2.00$ m on the x axis. Evaluate your results to five significant figures. (a) What is the total charge Q on the line? (b) Estimate the electric field at point P on the y axis at $y = 120$ m by assuming the entire charge Q is located at the origin. (c) Compare this estimate with the actual electric field at P due to the line charge. (d) Estimate the electric field at point P' on the y axis at $y = 2.00$ cm by assuming the line charge is infinitely long with the same linear charge density. (e) Compare this estimate with the actual electric field at P' due to the line charge. Use $k = 8.98755 \times 10^9$ N $\cdot$ m^2/C^2.

7. A sphere of radius a has uniform volume charge density ρ. Consider the Gaussian surface shown in the figure, consisting of a circular disk and a hemisphere, both of radius b, that are concentric with the charged sphere and together form a closed surface. What is the flux out of each segment of the Gaussian surface—that is, what is the flux out of the disk and what is the flux out of the hemisphere? Show that the net flux out of the Gaussian surface equals the charge inside it divided by ϵ_0.

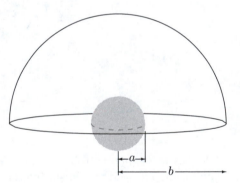

Chapter 23

Electric Potential

23.1 Potential Difference

In a Nutshell

The electric force is a conservative force, which means it has a scalar potential energy function associated with it. The change in electrical potential energy as a charge is moved in an electric field is found in the usual way, substituting the specific form for the electric force: $dU = -\vec{F} \cdot d\vec{\ell} = -q\vec{E} \cdot d\vec{\ell}$. The change in electrical potential energy per unit charge as that charge is moved from one point to another is defined to be the **potential difference** $dV = dU/q$ between those points.

The function V is called the **electric potential**, which is commonly shortened to simply **potential**. Like potential energy functions, this function is a scalar. Also like potential energy, it is only *changes* in electric potential that are important. The location of the zero point of the electric potential is arbitrary. For convenience, we often define the zero point of the electrical potential energy and the electric potential to be at the same location, but this is by no means required.

The scalar electric potential field V is continuous everywhere in space except at points where the electric field is infinite (the locations of point or line charges).

The SI unit of electric potential is the **volt** V. It is defined as a joule per coulomb: $1 \text{ V} = 1 \text{ J/C}$. The potential difference between two points is often referred to as the **voltage** between the two points. Voltage is measured with a voltmeter.

The definition of electric potential also reveals that the electric potential has dimensions of electric field times distance. Rearranging this relationship, we can find a new set of units for the electric field—one volt per meter. There is only one electric field, so the two sets of units must be equivalent: $1 \text{ N/C} = 1 \text{ V/m}$. So we may think of electric field as either a force per unit charge or a rate of change of electric potential with respect to distance in a given direction.

We often have particles like electrons or protons, with a charge of $\pm e$, that experience a change in potential difference. This leads us to define a very convenient unit of energy, the **electron volt** eV, which is simply the fundamental charge multiplied by a change in electric potential.

As positive charges accelerate in an electric field, they gain kinetic energy, and so must lose electrical potential energy. Since $\Delta U = q\,\Delta V$, the charge also accelerates toward a region of lower electric potential. Since positive charges accelerate in the direction of $\vec{E}$, $\vec{E}$ must point in the direction in which the potential V decreases most rapidly.

The idea of electric potential is quite useful for several reasons.
1. Electric potential is a scalar, and so is generally easier to calculate than electric fields.
2. Once the electric potential is known in a region of space it can be used to calculate the vector electric field.
3. Differences in electric potential are much easier to measure than electric fields.
4. Electric potential is directly related to electrical potential energy.

Physical Quantities and Their Units

Potential energy U	SI units of joules (J)
Electric potential V	SI units of volts (1 V = 1 J/C)
Electric field $\vec{E}$	SI units of newtons per coulomb (N/C) or volts per meter (V/m)
Electron volt eV	$1 \text{ eV} = 1.60 \times 10^{-19} \text{ C} \cdot \text{V} = 1.60 \times 10^{-19} \text{ J}$

Fundamental Equations

Differential change in potential energy	$dU = -\vec{F} \cdot d\vec{\ell}$
Definition of potential difference	$dV = \dfrac{dU}{q_0} = -\vec{E} \cdot d\vec{\ell}$
Definition of finite potential difference	$\Delta V = V_b - V_a = \dfrac{\Delta U}{q_0} = -\displaystyle\int_a^b \vec{E} \cdot d\vec{\ell}$

Common Pitfalls

> Remember that it is *changes* in electric potential that matter, not the absolute value of the potential at a given point. People often refer to the voltage at a given point. When we do this, we are implicitly referring to "the voltage at a given point with respect to the zero location." Clearly the zero location of the potential must be defined in some fashion.

> Remember that although electrical potential energy and potential often have the same zero point, they are not required to.

> Potential is a scalar quantity; it has no direction. Don't confuse it with the electric field, which is a vector; they are different quantities.

> The electrostatic force is always in the direction of decreasing potential energy, whereas the electric field is always in the direction of decreasing potential. These directions are not necessarily the same because the electrostatic force on a negative charge is in a direction opposite that of the electric field.

> The electrostatic force on a point charge is in the direction of the electric field only if the charge is positive.

> Be careful of the many uses of the letter "V." It is perfectly legal to say something like "$V = 10$ V." Why? The V on the left side of the expression is the variable we typically use to represent potential. The V on the right side represents the unit of volts. In print you should be able to distinguish them because the variable always appears italicized, V, and the unit is always upright, V. However, when you are doing problems, it can be much more difficult to distinguish between the two, so you have to pay attention to the context.

> To find the potential difference between two points from the electric field you must know the electric field everywhere along the path you take from point a to point b. If the electric field is unknown at any point along the path you choose, then the integral cannot be evaluated.

1. TRUE or FALSE: The volt is a unit of power.

2. An electron is released from rest in an electric field. Will it accelerate in the direction of increasing or decreasing potential? Why?

Try It Yourself #1

A uniform electric field of 2.00 kN/C is in the $+x$ direction. (a) What is the potential difference $V_b - V_a$ when point a is at $x = -30.0$ cm and point b is at $x = +50.0$ cm? (b) A test charge $q_0 = +2.00$ nC is released from rest at point a. What is its kinetic energy when it passes through point b?

Picture: Use the relationship between potential and electric field to calculate the potential difference. The change in the charge's potential energy, which can be calculated from the change in potential, will result in a change in the charge's kinetic energy.

Solve:

Sketch the physical situation described.	
Calculate the potential difference by relating it to the electric field.	$V_b - V_a = -1.60 \times 10^3$ V
Mechanical energy is conserved because the only force acting is the conservative electrical force. Since the initial kinetic energy is zero, the change in kinetic energy will be equivalent to the final kinetic energy. Find the change in potential energy and relate that to the change in kinetic energy.	$K_{final} = 3.2 \times 10^{-6}$ J

Check: The electric field is in the $+x$ direction, and point b is at a more positive x location. Since the electric field points in the direction of decreasing potential, we expect point b to have a lower potential than point a, which means the potential difference should be negative, which is what we found.

Taking It Further: If a negative charge instead of a positive charge were used in this problem, qualitatively how would your answers change?

Try It Yourself #2

An electric field is given by $E = (-3.00 \text{ kV/m}^2)x\,\hat{\imath}$ on the x axis. (a) What is the potential difference $V_b - V_a$ when point a is at $x = -30.0$ cm and point b is at $+50.0$ cm? (b) How much work is done by an external agent in bringing a test charge $q_0 = +2.00$ nC from rest at a to rest at b?

Picture: Use the relationship between potential and electric field to calculate the potential difference. The work done by the external agent plus the work done by the electric field equals the change in the charge's kinetic energy.

Solve:

Sketch the physical situation described.	
Calculate the change in electric potential due to the electric field as the charge moves from point a to point b.	$\Delta V = 240$ V
Determine the change in kinetic energy of the particle.	
Determine the work done on the particle by the electric field.	
Apply the work–energy theorem to determine the work done by the external agent.	$W_{\text{ext}} = +4.80 \times 10^{-7}$ J

Check: Since the change in kinetic energy is zero, there should be no net work done on the test charge. Therefore the work done by the electric field and the work done by the external agent should have the same magnitude but opposite signs.

Taking It Further: If point b were moved to $x = +30.0$ cm, how much work would be required from the external agent? Explain, by describing the physical system, how this is possible.

23.2   Potential Due to a System of Point Charges

In a Nutshell

By applying the definition of potential difference to a point charge, we can derive the electric potential function for a point charge, $V = \dfrac{kq}{r} + V_0$. It is most convenient to let the integration constant $V_0 = 0$, which gives rise to the **Coulomb potential** $V = kq/r$. This function implicitly defines $V = 0$ to occur infinitely far away from the point charge creating the electric field and electric potential.

The potential energy function for two point charges can then be shown to be $U = kqq'/r$, again letting the potential energy $U = 0$ when the charges are separated by an infinite distance.

Because the electric field obeys the principle of superposition and electric potential is calculated from the electric field, electric potential also obeys the principle of superposition—that is, to find the electric potential at some point in space due to a distribution of charges you calculate the potential at that point in space due to each charge individually then add them together.

Calculating V Due to a System of Point Charges

Picture: We can use $V = \sum_i \dfrac{kq_i}{r_i}$ to calculate the potential at a field point due to any collection of point charges if each point charge is a finite distance from every other point charge.

Solve:

1. Sketch the charge configuration and include suitable coordinate axes. Label each point charge with a distinct symbol, such as q_1, q_2, ... Draw a straight line from each point charge q_i to the field point P and label it with a suitable symbol such as r_{1P}, r_{2P}, ... A careful drawing can be very helpful in relating the distances of interest to the distances given in the problem statement.

2. Use the formula $V = \sum_i \dfrac{kq_i}{r_i}$ to calculate the potential at P due to the presence of the point charges.

Check: If the field point is arbitrarily chosen, take the limit as the field point goes to infinity. In that limit, the potential must approach zero.

Important Derived Results

Coulomb potential of a point charge	$V = \dfrac{kq}{r}$ if $V = 0$ as $r \to \infty$
Potential energy of two point charges	$U = \dfrac{kqq'}{r}$ if $U = 0$ as $r \to \infty$
Potential due to a system of point charges	$V = \sum_i \dfrac{kq_i}{r_i}$

Common Pitfalls

> ➤ Remember that electrical potential energy can be either positive or negative. If the Coulomb potential is used, then a negative potential energy means the two charges involved are attracted to each other, while a positive potential energy means the two charges are repelled from each other.

3. TRUE or FALSE: Charges of like sign have more electrostatic potential energy when they are farther apart.

4. Is the electric potential at a location P associated with a charge at P or the location itself? Explain.

Try It Yourself #3

Consider a 1.00-m^3 cube that has +2.00-μC charges located at seven of its corners as shown. (a) Find the potential at the vacant corner. (b) How much work by an external agent is required to bring an additional +2.00-μC charge from rest at infinity to rest at the vacant corner?

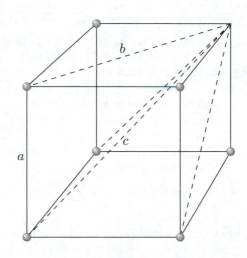

Picture: The potential at the remaining point is the sum of the potentials due to each of the individual charges. By the work–energy theorem, the work done by an external force will equal the sum of the change in kinetic and potential energy of the charge.

Solve:

Examine the geometry of the cube to find the distances b and c *algebraically* in terms of the length of one side of the cube, a.	

The total potential at the remaining corner is the sum of the potential from each of the charges. As can be seen in the figure, there are three charges a distance a from the vacant corner, three charges a distance of b from the vacant corner, and one charge a distance of c from the corner. Find the total potential at the vacant corner.	$V = 1.02 \times 10^5$ V
The work done by an external agent will equal the total change in energy of the external particle as it moves from infinitely far away to the remaining corner. The change in potential energy is related to the change in electric potential experienced by the charge as it moves to the remaining corner.	$W_{\text{ext}} = 0.204$ J

Check: Since all the charges are positive, they will repel the last charge, so the external agent should have to do positive work, which is what we found.

Taking It Further: How does this problem change, if at all, if a -2.00-μC charge is brought in to fill the empty corner rather than a positive charge?

Try It Yourself #4

A charge of $+2.00$ μC is at the origin and a charge of -3.00 μC is on the y axis at $y = 40.0$ cm. (a) What is the potential at point a, which is on the x axis at $x = 40.0$ cm? (b) What is the potential difference $V_b - V_a$ when point b is at $(40.0 \text{ cm}, 30.0 \text{ cm})$?

Picture: The potential at a point is the algebraic sum of the potentials due to each of the individual charges. Since we have point charges, we let $V = 0$ infinitely far from the charge distribution.

Solve:

Sketch the physical situation described. Label each charge and draw and label a line from each charge to the two points a and b.	
Find the distances from each charge to point a. Make sure to label them properly.	
Use the definition of electric potential of a point charge to find the total electric potential at point a due to the two charges.	$V_a = -2730$ V
Find the distances from each charge to point b. Make sure to label them properly.	
Use the definition of electric potential of a point charge to find the total electric potential at point b due to the two charges.	
Find the potential difference $V_b - V_a$.	$V_b - V_a = -26\,700$ V

Check: Point b is closer to the negative charge, so we would expect its potential to be more negative than that at point a.

Taking It Further: How much work is required to move an electron at rest from point a to rest at point b?

23.3 Computing the Electric Field from the Potential

In a Nutshell

By rearranging the expression $dV = -\vec{E} \cdot d\vec{\ell}$, we can solve for the electric field: $E_{\text{tan}} = -dV/d\ell$, where E_{tan} is the component of the electric field parallel to $d\vec{\ell}$. When the displacement $d\vec{\ell}$ is perpendicular to the electric field then the change in potential $\Delta V = 0$. The maximum increase in V occurs when $d\vec{\ell}$ is in the same direction as $-\vec{E}$. Mathematically, we say that $\vec{E}$ is the negative **gradient** of the electric potential function.

Important Derived Results

Electric field from the potential function

$$E_x = -\frac{\partial V}{\partial x}$$
$$E_y = -\frac{\partial V}{\partial y}$$
$$E_z = -\frac{\partial V}{\partial z}$$
$$E_r = -\frac{\partial V}{\partial r}$$
$$\vec{E} = -\vec{\nabla} V$$

Common Pitfalls

> To find the local electric field from the potential, you need to know the spatial derivatives of the potential only at that point, not everywhere in space.

5. TRUE or FALSE: The electric field equals the *negative* of the gradient of the potential.

6. In a region where there is an electric field, two nearby points are at the same potential. What is the angle between the line joining the points and the direction of the electric field? Explain.

Try It Yourself #5

The electric potential has the form $V = (3.00\ \text{V/m}^2)x^2 - (4.00\ \text{V/m})y - (8.00\ \text{V/m}^3)z^3$. Find the electric field at the point $(2.00\ \text{cm}, -6.00\ \text{cm}, -10.0\ \text{cm})$.

Picture: Use the fact that the electric field is the negative gradient of the potential to find the full vector electric field.

Solve:

Find the x component of the electric field by taking the negative derivative of the potential with respect to x and evaluating at the specified point.	
Find the y component of the electric field by taking the negative derivative of the potential with respect to y and evaluating at the specified point.	
Find the z component of the electric field by taking the negative derivative of the potential with respect to z and evaluating at the specified point.	
Find the total electric field.	$\vec{E} = -(0.120\ \text{V/m})\hat{\imath} + (4.00\ \text{V/m})\hat{\jmath}$ $+ (0.240\ \text{V/m})\hat{k}$

Check: The units check out properly.

Taking It Further: In what direction will be the initial acceleration of a doubly ionized sodium ion located at that position?

Try It Yourself #6

A charge $q_1 = +2q$ is at the origin and a charge $q_2 = -q$ is on the x axis at $x = a$. (a) Find an expression for the electric potential $V(x)$ on the x axis in the region for $0 < x < a$. (b) Use your result from part (a) to find an expression for the electric field $\vec{E}$ in the same region.

Picture: The potential at a point is the algebraic sum of the potentials due to each of the individual charges. The electric field is the negative of the gradient of the potential.

Solve:

Draw a sketch of the two charges on the x axis.	
Determine the potential at an arbitrary coordinate x due to the charge at the origin.	
Determine the potential at an arbitrary coordinate x due to the charge at $x = a$. Watch your signs, and be careful when constructing an expression for the distance from the charge to the point. Make sure the distance gets smaller as the point gets closer to the charge at $x = a$.	
Sum the potentials of the previous two steps to find the potential at this arbitrary location x in the region $0 < x < a$.	$V = kq\left(\dfrac{2}{x} - \dfrac{1}{a-x}\right)$
The electric field in the same region is the negative gradient of the potential.	$\vec{E} = kq\left(\dfrac{2}{x^2} + \dfrac{1}{(a-x)^2}\right)\hat{\imath}$

Check: The electric field should point from the positive charge toward the negative charge, which it does.

Taking It Further: The electric field in this region is always in the $+x$ direction. Is the potential ever zero? If so, where, and why is this possible? If not, why is it not possible?

23.4   Calculations of V for Continuous Charge Distributions

In a Nutshell

For a continuous charge distribution the potential V at a point is equal to the sum of the potentials due to all elements of charge dq. Thus, for a charge distribution that is confined to a finite region of space, the potential is $V = \int_Q \dfrac{k\,dq}{r}$.

Calculating the potential for charge distributions that are *not* confined to a finite region of space (such as a uniformly charged infinite plane) requires a different approach. For such charge distributions the potential cannot be zero at infinity. To calculate the potential for these distributions, we first find the electric field and then use the relation $V - V_0 = -\displaystyle\int_{P_0}^{P} \vec{E} \cdot d\vec{\ell}$, where P_0 is the reference point at which the potential is V_0 and P is the field point. This completely general approach works for any charge distribution.

The relationship $V - V_0 = -\displaystyle\int_{P_0}^{P} \vec{E} \cdot d\vec{\ell}$ can be used to find potential functions for a wide variety of geometries. The details are provided in the text, which you should consult. Some of the end results are given below in Important Derived Results.

Fundamental Equations

Potential due to a continuous, finite charge distribution $V = \displaystyle\int \dfrac{k\,dq}{r}$

Important Derived Results

Potential on the axis of a charged ring of radius a $V = \dfrac{kQ}{\sqrt{z^2 + a^2}}$

Potential on the axis of a uniformly charged disk of radius R $V = 2\pi k\sigma|z|\left(\sqrt{1 + \dfrac{R^2}{z^2}} - 1\right)$

Potential near an infinite plane of charge $V = V_0 - 2\pi k\sigma|x|$

Potential of a spherical shell of charge with radius R $V = \begin{cases} \dfrac{kQ}{r} & (r \geq R) \\[2mm] \dfrac{kQ}{R} & (r \leq R) \end{cases}$

Potential due to an infinite line charge $V = 2k\lambda \ln \dfrac{R_{\text{ref}}}{R}$

Common Pitfalls

> ➤ When calculating the electric potential due to an infinite charge distribution, you *cannot* integrate the potential from each differential charge element. You must first find the electric field and calculate the potential from it.

7. TRUE or FALSE: If the expression used for the potential of a point charge dq is $dV = k\, dq/r$, this potential must approach zero as r approaches infinity.

8. If the electric field is zero throughout some volume of space, what can you say about the potential in that volume? Why?

Try It Yourself #7

The x axis coincides with the symmetry axis of a uniformly charged thin disk with radius R and uniform positive surface charge density σ centered at the origin. (a) Make a sketch of E_x versus x for $-4R < x < +4R$. (b) Make a sketch of $V(x)$ versus x for $-4R < x < +4R$.

Picture: Use the expressions for the electric field and potential on the axis of a uniformly charged disk.

Solve:

Make a sketch of the situation.	
Recall the equation for the magnitude E of the electric field on the axis of a uniformly charged disk.	
Sketch $E_x(x)$. The electric field is directed away from the disk on both sides of it. Thus E_x is positive for $x > 0$ and negative for $x < 0$.	

Recall the expression for the potential on the axis of a uniformly charged disk.	
Sketch $V(x)$. This function is continuous and always positive.	

Check: We know that $E_x = -\partial V/\partial x$. Do your graphs share this relationship?

Taking It Further: How do your graphs change if the disk has an infinite radius? Sketch them here.

Try It Yourself #8

A uniformly charged sphere has radius a and total charge Q. (a) Find the electric field E_r for $0 < r < \infty$. (b) Use your result from part (a) to find an expression for the electric potential $V(r)$ everywhere.

Picture: Use Gauss's law to find the electric field in all regions of space. Then use $\Delta V = -\int \vec{E} \cdot d\vec{\ell}$ to find the potential. Since the sphere is finite in size, we can let $V = 0$ as $r \to \infty$.

Solve:

Use Gauss's law to find the electric field due to the charged sphere everywhere in space. You may want to refer to Chapter 22 for the steps involved.	$$E_r = \begin{cases} \dfrac{kQ}{r^2} & (r \geq a) \\[2mm] \dfrac{kQr}{a^3} & (r \leq a) \end{cases}$$
Use the relationship between $\vec{E}$ and ΔV outside the sphere, as well as letting $V = 0$ as $r \to \infty$ to find the electric potential outside the sphere.	$V = kQ/r$ for $r \geq a$
Use the above expression to determine the potential right at the surface of the charged sphere.	
Using the potential at the surface of the sphere as your reference potential, now use the relationship between $\vec{E}$ inside the sphere and ΔV to find an expression for V inside the sphere.	$V = \dfrac{1}{2}\dfrac{kQ}{a}\left[3 - \left(\dfrac{r}{a}\right)^2\right]$ for $r < a$

Check: Outside the sphere, the field and potential should both look like those from a point charge centered on the sphere and with total charge Q. The electric potential should also be continuous. Is it?

Taking It Further: Plot both E_r and V as functions of r.

23.5   Equipotential Surfaces

In a Nutshell

All points on any surface that is perpendicular to electric field lines are at the same potential. We call such surfaces **equipotential surfaces**. When a charge moves on such a surface, the electric field does no work on it. The entire body of a conductor in electrostatic equilibrium is an equipotential volume (or just an "equipotential"). The equipotential surfaces are shown by the dotted lines, and the solid lines with arrows show the electric field lines between a $+2q$ charge on the bottom and a $-q$ charge on the top.

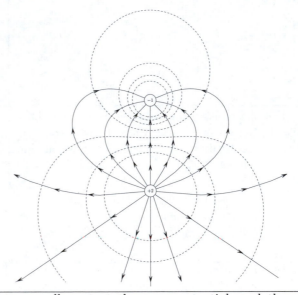

Conductors not in contact with one another are usually not at the same potential, and the potential difference between them depends on their sizes, shapes, locations, relative orientation, and charges.

When conductors are brought into contact, charge is transferred between them as the charge distributions rearrange to reestablish electrostatic equilibrium. In equilibrium, conductors touching each other constitute a single equipotential region. Consider the special case of a conductor that is completely enclosed by a larger conductor. If the smaller conductor touches the inside surface of the outer conductor, all the charge that was on the inner conductor flows to the outer surface of the outer conductor. Repetition of this process is what charges a Van de Graaf generator to a high potential.

The total amount of charge that can be transferred to a conductor is limited only by **dielectric breakdown** of the insulating material surrounding the outer surface of the outer conductor. Dielectric breakdown occurs when an electric field is strong enough to accelerate electrons or ions sufficiently to ionize the molecules with which they collide. Dielectric breakdown occurs in dry air when the magnitude of the electric field reaches $E_{\max} \approx 3 \times 10^6$ V/m. The electric field strength at which dielectric breakdown occurs for a particular material is called the **dielectric strength** of the material.

The electric discharge through the air resulting from dielectric breakdown is called **arc discharge**. Charge tends to concentrate on sharp points or corners on the surface of a conductor where the radius of curvature is very small. Thus an electric field is strongest just outside the regions of the surface of a conductor where the radius of curvature of the surface is least; so dielectric breakdown occurs most frequently at sharply pointed regions on the surface of a conductor.

Physical Quantities and Their Units

Dielectric strength of air $\qquad E_{\max} \approx 3 \times 10^6$ V/m

Common Pitfalls

> The electric potential in a conductor material in static equilibrium is constant, but not necessarily zero.

9. TRUE or FALSE: Dielectric breakdown in air occurs when the electric field is strong enough to ionize the air molecules.

10. If conductors are equipotential surfaces, why does dielectric breakdown occur preferentially at a sharp corner of the conductor rather than on a flatter portion of the conductor?

Try It Yourself #9

An isolated conducting sphere is to be charged to 100 kV. What is the smallest radius the sphere can have if its electric field is not to exceed the dielectric strength of air?

Picture: All the charge on a conducting sphere resides on its surface. Obtain expressions for both the electric field strength and electric potential. Use these expressions to find an expression for the electric field of a charged conducting sphere in terms of its potential and radius. Solve for R.

Solve:

Write the expression for the electric field outside a uniformly charged shell. The electric field is greatest right at the surface of the shell.	
Do the same for the electric potential just at the surface of the sphere.	

Solve the result of the previous step for kQ, substitute into the result from the first step, and solve for R. Use the dielectric breakdown strength of air for E.	
	$R = 3.33$ cm

Check: The units work out correctly.

Taking It Further: If the sphere were to be embedded in glass, with a dielectric strength of approximately 14×10^6 V/m, would the minimum radius of the sphere increase or decrease? Explain.

Try It Yourself #10

Find the maximum surface charge density σ_{max} that a planar conducting surface in air can sustain.

Picture: Use Gauss's law to find the electric field of a planar surface. Set that electric field equal to the breakdown field of air, and solve for σ_{max}.

Solve:

Draw a sketch of a segment of a conducting surface.	
Use Gauss's law to obtain an expression for the electric field just outside the surface in terms of the surface charge density. You may wish to refer to Chapter 22.	

The maximum electric field strength equals the dielectric strength of air. Use this to solve for σ_{max}.	
	$\sigma_{max} = 2.66 \times 10^{-5}$ C/m^2

Check: The units work out properly.

23.6 Electrostatic Potential Energy

In a Nutshell

The **electrostatic potential energy** of a system of point charges is the work needed to bring the charges from an infinite separation to their final positions.

To calculate the electrostatic potential energy of a system of point charges, we start with one of the charges in its final position and then calculate the work required to bring a second charge from infinity to its final position. Then we calculate the work required to bring a third charge from infinity to its final position in the resultant electric field of the first two charges, and so forth. The result is that the potential energy of the system can be written as $U = \dfrac{1}{2} \sum_{i=1}^{n} q_i V_i$, where q_i is the charge of the ith point charge and V_i is the potential at the location of the ith charge due to the presence of the other charges in the distribution.

If instead of point charges we talk about the potential energy required to bring n conductors each with a charge Q_i and potential V_i the electrostatic potential energy becomes $U = \dfrac{1}{2} \sum_{i=1}^{n} Q_i V_i$.

Important Derived Results

Electrostatic potential energy of a system of point charges $\qquad U = \dfrac{1}{2} \sum_{i=1}^{n} q_i V_i$

Electrostatic potential energy of a system of conductors $\qquad U = \dfrac{1}{2} \sum_{i=1}^{n} Q_i V_i$

Common Pitfalls

> The electrostatic potential energy of a single, isolated point charge is zero because it takes no work to place the first charge of a charge distribution.
> Make sure you do not "double-count" when adding the total potential energy of system. Each pair of charges must occur only once.

11. TRUE or FALSE: Charges of opposite sign have more electrostatic potential energy when they are farther apart.

12. Explain the significance of the factor of $\frac{1}{2}$ in the two Important Derived Results.

Try It Yourself #11

As shown in the figure, three particles, each
with charge q, are at different corners of a
rhombus with sides of length a and with one
diagonal of length a and the other of length b.
(a) What is the electrostatic potential energy
of this charge distribution? (b) How much
work by an external agent is required to bring
a fourth particle, also of charge q, from rest
at infinity to rest at the vacant corner of the
rhombus? (c) What is the total electrostatic
potential energy of the four charges?

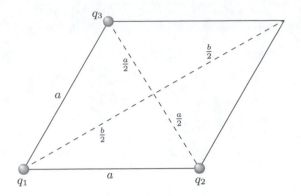

Picture: The electrostatic potential energy is the work by an external agent required to assemble
the charge distribution. Determine the work required to bring in each charge separately, and sum
the result.

Solve:

If we start with empty space, the work required to bring q_1 into position and the potential energy of q_1 are zero.	
Calculate the change in potential energy required to bring charge q_2 into place while q_1 is present.	
Calculate the change in potential energy that results from bringing charge q_3 into place in the presence of q_1 and q_2. The final resting place of q_2 is a distance a from each of the other charges.	
The total potential energy of the system is the sum of the potential energies. Remember $q_1 = q_2 = q_3 = q$.	$U = \dfrac{3kq^2}{a}$

To find the work required to bring in the fourth charge, we can first find the electric potential at the fourth corner, which is the sum of the electric potential due to the other three charges.	
Use the electric potential of the previous step to calculate the work required by an external agent to bring the fourth charge into place.	
The total potential energy of the four charges is the sum of the potential energy of the three charges plus the work required to bring in the fourth charge.	$$U_{\text{total}} = \left(5 + \frac{1}{\sqrt{3}}\right)\frac{kq^2}{a}$$

Check: Make sure the units of each expression will give you joules if you use the SI system of units.

Try It Yourself #12

Three $+2.00$-μC charges are located on the x axis at $x = 0.00$, $x = 10.0$ cm, and $x = 20.0$ cm. (a) What is the electrostatic potential energy of this distribution? (b) Each of the three charges is replaced by a -2.00-μC charge. Now what is the electrostatic potential energy of the distribution?

Picture: The electrostatic potential energy is the work by an external agent required to assemble the charge distribution.

Solve:

Sketch the three charges on the axis.	

Determine the work required to bring the first charge at $x = 0$ into place.	
Determine the work required to bring the second charge at $x = 10.0$ cm into place in the presence of the first charge.	
Determine the work required to bring the final charge into place in the presence of the first two charges.	
The total potential energy is the sum of the work calculated in the previous steps.	$U = +0.899$ J
Repeat the second through fifth steps if the charges are negative instead of positive.	$U = +0.899$ J

Check: The units all work out.

Taking It Further: Explain why the potential energy is the same regardless of whether all positive or all negative charges are used.

QUIZ

1. TRUE or FALSE: Electric field lines point in the direction of decreasing potential.

2. TRUE or FALSE: The term *voltage* refers to a potential difference.

3. Explain why electric potential requires the existence of only one charge, but a finite electrical potential energy requires the existence of two charges.

4. Does it make sense to say that the voltage at some point in space is 10.3 V? Why or why not?

5. Explain how an electron will accelerate toward a region of lower electric potential energy but higher electric potential.

6. An infinitely long line charge of linear charge density $\lambda = 2.00$ nC/m lies on the z axis. (a) Find the potential difference $V_c - V_b$ from the electric field when point b is at $(40.0$ cm, 30.0 cm, $0)$ and point c is at $(200$ cm, $0, 50.0$ cm$)$. Does your answer depend on your choice of the reference point at which $V = 0$? (b) Describe the equipotential surfaces.

7. A charged semicircular ring with radius $r = 7.00$ cm centered at the origin has a linear charge density that varies with angle according to $\lambda = (4.00\ \mu\text{C/m})\sin\theta$. Using direct integration determine the electric potential at the origin due to this charge distribution.

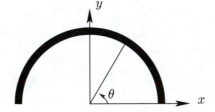

Chapter 24

Capacitance

24.1 Capacitance

In a Nutshell

A **capacitor** is a device with two conductors insulated from each other. We will see in the next section how such a device is used to store charge. One conductor of the capacitor carries a charge $+Q$ and the other a charge $-Q$. The ratio of the magnitude of the charge Q to the magnitude of the potential difference V between the conductors is called the **capacitance** of the capacitor. The capacitance is always a positive quantity. The SI unit of capacitance is the farad F, which is equal to one coulomb per volt.

Capacitance is a measure of the capacity of a device to store charge with a given potential difference between the electrodes. Although it appears to depend on both Q and V, these two quantities are proportional. As a result the capacitance depends on the geometry of the device and the insulating material between the conductors.

Calculating Capacitance

Picture: Make a sketch of the capacitor that has a charge of $+Q$ on one conductor and a charge of $-Q$ on the other conductor.

Solve:

1. Determine the electric field $\vec{E}$ between the conductors (this is usually done by using Gauss's Law).
2. Determine the magnitude of the potential difference V between the two conductors by integrating $dV = -\vec{E} \cdot d\vec{\ell}$.
3. The capacitance is equal to $C = Q/V$.

Check: Check that the result depends only on constants, the permittivity of the insulator and on geometric factors such as lengths and areas.

The text uses the above strategy to solve for the capacitance of several geometries. Although the resulting expressions are listed in Important Derived Results, you should refer to the text for the meaning of the variables in those expressions.

Physical Quantities and Their Units

Capacitance C	$1\,\text{F} = 1\,\text{C/V}$
Permittivity of free space	$\epsilon_0 = 8.85 \times 10^{-12}\,\text{C}^2/(\text{N} \cdot \text{m}^2) = 8.85 \times 10^{-12}\,\text{F/m}$

Fundamental Equations

Definition of capacitance	$C = \dfrac{Q}{V}$

Important Derived Results

Capacitance of an air-filled parallel-plate capacitor $\qquad C = \dfrac{\epsilon_0 A}{d}$

Capacitance of an air-filled long cylindrical capacitor $\qquad C = \dfrac{2\pi\epsilon_0 L}{\ln(R_2/R_2)}$

Common Pitfalls

> ➤ The charge of a capacitor refers to the magnitude of the charge on one of the plates. The total charge on the two plates is, of course, zero.
> ➤ Although we simply use V in the expression for the capacitance, remember that potentials are always potential differences. What we mean by V is the potential difference between the two conductors of the capacitor.
> ➤ Remember that capacitance is *defined* as $C = Q/V$, but it depends on neither of those quantities. It depends only on the geometry and the insulating material of the capacitor.
> ➤ Capacitance is always a positive quantity.

1. TRUE or FALSE: The capacitance of a capacitor is the maximum charge it can hold.

2. Explain, defending all statements that you make, why the capacitance depends neither on the stored charge Q nor on the potential difference V between the conductors.

Try It Yourself #1

A parallel-plate, air-gap capacitor has a charge of 20.0 μC and a gap width of 0.100 mm. The potential difference between the plates is 200 V. (a) What is the electric field in the region between the plates? (b) What is the surface charge density on the positive plate?

Picture: The electric field between the plates of a parallel-plate capacitor is uniform. The surface charge density of each plate contributes equally to the electric field.

Solve:

Use the relationship among potential, electric field, and distance to find the electric field, remembering that the electric field is uniform in a parallel-plate capacitor.	
	$E = 2.00 \times 10^6$ V/m
The electric field between the plates is due to two thin sheets of charge, one on each of the plates. Recall the electric field due to a sheet of charge, and use the superposition principle to find the surface charge density.	
	$\sigma = 1.77 \times 10^{-5}$ C/m^2

Check: These values seem reasonable and the units work out properly.

Taking It Further: If the plates of this capacitor are moved closer together while the *charge* remains constant, how are the electric field, surface charge density, and potential difference going to change, if at all? Explain.

Try It Yourself #2

Find an expression for the capacitance of a concentric, spherical, air-filled capacitor with inner conductor of radius R_1 and outer radius R_2.

Picture: Follow the Calculating Capacitance problem-solving strategy outlined above.

Solve:

Sketch a cross section of the two spheres. Indicate a positive charge $+Q$ on the inner sphere and an equal-magnitude opposite charge $-Q$ on the outer sphere.	
Use Gauss's law to determine the electric field $\vec{E}$ between the two conductors. You will need to use a Gaussian surface of some arbitrary radius $R_1 < r < R_2$. Remember that it is only the enclosed charge that matters.	

Integrate $dV = -\vec{E} \cdot d\vec{\ell}$ to find the potential difference between the spheres. The radius r should be your integration variable. Watch out for your signs, but remember in the end it is the magnitude of the potential difference that is needed.	
Substitute values into the definition of capacitance to find the capacitance of these two spheres.	$C = 4\pi\epsilon_0 \dfrac{R_1 R_2}{R_2 - R_1}$

Check: The result is positive and depends only on geometric factors and the permittivity of free space.

Taking It Further: If the inner sphere is made larger, will the capacitor store more or less charge per volt? Why?

24.2   The Storage of Electrical Energy

In a Nutshell

The energy stored in a capacitor is equal to the work required to separate the charges on the conductors of the capacitor. A careful analysis (see page 807 of the text) results in three equivalent expressions for the electrical potential energy stored in a capacitor: $U = Q^2/(2C) = QV/2 = CV^2/2$.

We can think of the electrical potential energy as the work required to establish the electric field between the conductors of a capacitor. In other words, we can think of the energy as being stored in the electric field itself. This is an entirely general result. Wherever an electric field is present in space, there is some **energy density** (energy per unit volume) $u_e = \frac{1}{2}\epsilon_0 E^2$ stored in that electric field.

Important Derived Results

Energy stored in a capacitor

$$U = \frac{1}{2}\frac{Q^2}{C} = \frac{1}{2}QV = \frac{1}{2}CV^2$$

Energy density of an electrostatic field

$$u_e = \frac{1}{2}\epsilon_0 E^2$$

Common Pitfalls

> ➤ It is common to want to know how the energy stored in a capacitor changes when some other quantity changes. For these questions it is imperative that you identify which quantity remains constant: C, Q, or V. Because these three quantities are intertwined, only one can remain constant. There will always be two quantities that change. Although any form of the expression for the energy stored in a capacitor can be used to determine how the energy will change, it is always easiest to choose the form that contains (a) the quantity you have determined must remain *constant* and (b) the quantity that you are explicitly told is *changing*.
> ➤ If a capacitor is attached to a battery, the potential difference is constant. If a capacitor is *isolated* then the charge remains constant because there is nowhere for it to go. If the geometry remains fixed, then the capacitance is the constant quantity.

3. TRUE or FALSE: The electrostatic potential energy stored in a capacitor with charge Q and potential difference V is the product QV.

4. Explain the fundamental origin of the factor of $\frac{1}{2}$ in the expressions for the energy stored in a capacitor.

Try It Yourself #3

An air-gap, parallel-plate capacitor with area A and gap width d is connected to a battery that maintains the plates at potential difference V. (a) Find expressions for the electric field and energy density in the region between the plates and for the charge on the positive plate. (b) Use the expression for energy density to obtain an expression for the electrostatic potential energy of the capacitor. (c) The plates are pulled apart, doubling the gap width, while they remain in electrical contact with the battery terminals. By what factor does the potential energy of the capacitor change?

Picture: Use the expressions developed in this chapter to find the requested quantities.

Solve:

Use the expression relating the potential difference to the electric field to find the electric field. Remember that for parallel-plate capacitors we treat the electric field as uniform.	$E = V/d$
Use the expression for energy density and the electric field found in the first step to find the energy density.	$u_e = \dfrac{\epsilon_0}{2}\left(\dfrac{V}{d}\right)^2$

Use the definition of capacitance and the capacitance of a parallel-plate capacitor to find the charge on the positive plate.	$$Q = \frac{\epsilon_0 A V}{d}$$
The total potential energy is the energy density multiplied by the volume of the capacitor.	$$U = \frac{\epsilon_0 A V^2}{2d}$$
Use the expression of the previous step to determine how the potential energy changes when the gap width is increased.	When d is doubled, U will be halved.

Check: The units of each expression work out properly.

Taking It Further: If the capacitor is removed from the battery and thus isolated, what happens to the stored potential energy when the gap width is doubled? Explain.

Try It Yourself #4

An air-filled parallel-plate capacitor is attached to a battery with a voltage V. While attached to the battery, the area of the plates is doubled and the separation of the plates is halved. During this process, what happens to (a) the capacitance, (b) the charged stored on the positive plate of the capacitor, (c) the potential across the plates of the capacitor, and (d) the potential energy stored in the capacitor, as compared to the original configuration?

Picture: Use the expressions for capacitance, the capacitance of a parallel-plate capacitor, and the energy stored in a capacitor to solve this problem.

Solve:

Use the expression for the capacitance of an air-filled parallel-plate capacitor to determine how the capacitance varies with plate separation and area, which you can use to find the change in capacitance.	The capacitance will increase by a factor of 4.
Use the definition of capacitance to determine how the stored charge will change. Substitute the expression for the plane parallel-plate capacitance to see the variation due to the area and separation. Since the capacitor remains attached to the battery, the potential will remain constant.	The stored charge will increase by a factor of 4.
Use the fact that the capacitor remains attached to the battery to determine the change in the potential.	The potential remains constant.
Since the capacitor remains attached to the battery, the potential will remain constant. We also know how the capacitance changes. So use the expression for the energy stored in a capacitor that contains the potential and the capacitance. Substitute in the expression for a plane parallel-plate capacitor to determine the change in energy.	The stored potential energy will increase by a factor of 4.

Check: The potential must remain constant since the capacitor remains attached to the battery.

Taking It Further: How would your answers above change if, once the capacitor was charged by the battery, it was disconnected from the battery while the area and separation were changed as above?

24.3   Capacitors, Batteries, and Circuits

In a Nutshell

A **battery** is a "charge pump". A chemical reaction inside the battery attempts to maintain a constant voltage, called the **open-circuit terminal voltage** between the positive terminal (anode) and the negative terminal (cathode). When charge is shared between the battery and a capacitor, for instance, the terminal voltage is temporarily reduced until the chemical reaction separates additional charge on the terminals.

Symbols used to represent a battery and a capacitor in a circuit diagram are shown at the right. The long line on the battery represents the positive terminal and the short line represents the negative terminal. Any number of pairs of these lines may appear on this symbol, including just one pair. The + and − symbols may or may not be present. The symbol for a capacitor is similar to that for a battery, with two *equal length* parallel lines. Other variations on this symbol do exist, the most common being the symbol for a polar capacitor, which is also shown.

Symbol for a battery

Symbol for a capacitor

Symbol for a polar capacitor

A **junction** is a point in a circuit where one wire divides into two or more wires. The circuits below show junctions represented by dots, to help you identify them. However, many circuits that you see (including several in the text and this study guide) will not mark each junction with a dot.

Kirchhoff's loop rule states, "The changes in potential around any closed path always sum to zero."

Capacitors are said to be **connected in parallel** if one plate of each capacitor is connected directly to one plate of every other capacitor *and* the second plate of each capacitor is directly attached to the second plate of every other capacitor. A group of capacitors connected in parallel can be represented by a single **equivalent capacitance**. The equivalent capacitor has the same capacitance as the combination, stores the same amount of charge as the combination, and has the same potential difference across its plates as the combination. The equivalent capacitance for capacitors in parallel is given by $C_{\text{eq},\parallel} = C_1 + C_2 + C_3 + \ldots$. Capacitors (or any circuit elements, for that matter) in parallel each have the same potential between their electrodes. The charge stored by the equivalent capacitance is the sum of the charges stored on each individual capacitor.

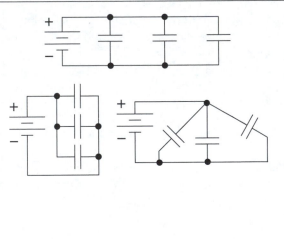

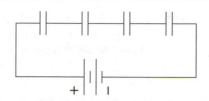

When capacitors are connected linearly, with the plate of one capacitor attached directly to a plate of the next capacitor, with *no junctions*, they are said to be attached in **series**. A group of capacitors connected in series can also be represented by a single **equivalent capacitance**. The equivalent capacitor has the same capacitance as the combination, stores the same amount of charge as the combination, and has the same potential difference across its plates as the combination. The equivalent capacitance for capacitors in series is given by $C_{eq,series}^{-1} = C_1^{-1} + C_2^{-1} + C_3^{-1} + \ldots$. Capacitors in series each store the same charge, and that charge is equal to the charge stored by the equivalent capacitance. The voltage across the equivalent capacitance is equal to the sum of the voltages across each individual capacitor connected in series.

Important Derived Results

Parallel capacitors equivalent capacitance

$$C_{eq,\parallel} = C_1 + C_2 + C_3 + \cdots = \sum C_i$$

Series capacitors equivalent capacitance

$$\frac{1}{C_{eq,series}} = \frac{1}{C_1} + \frac{1}{C_2} + \frac{1}{C_3} + \cdots = \sum \frac{1}{C_i}$$

Common Pitfalls

➤ When capacitors are attached in series, the total stored charge is *not* the sum of the individual charges. The total stored charge is equal to the charge on just one of the capacitors. It is the *voltages* that add.

➤ When capacitors are attached in parallel, the total voltage is the same as the voltage across each individual capacitor, but the charges add.

➤ Make sure the additive quantity—voltage for capacitors in series and charge for capacitors in parallel—sum to give the equivalent voltage or charge, respectively.

5. TRUE or FALSE: Capacitors in parallel necessarily have the same charge.

6. The capacitance of several capacitors in series is less than that of any of the individual capacitances. What, then, is the advantage of attaching several capacitors in series?

Try It Yourself #5

A 10.0-μF capacitor, a 40.0-μF capacitor, and a 100.0-μF capacitor are connected in series. A 12.0-V battery is connected across this combination. (a) What is the equivalent capacitance of the combination? (b) What is the charge on each capacitor? (c) What is the potential difference across each capacitor?

Picture: Use the expression for equivalent capacitance of capacitors in series. Capacitors in series all hold the same charge, and that charge is the same as the charge stored by the equivalent capacitance. Use the stored charge and the capacitance of each capacitor to find the potential difference across each capacitor.

Solve:

Sketch the schematic of capacitors to visualize the circuit.	
Determine the equivalent capacitance for capacitors in series.	$C_{\text{eq}} = 7.41 \ \mu\text{F}$
The charge stored by each capacitor is the same as the charge stored by an equivalent capacitance. Find the charge stored on the equivalent capacitor.	$Q = 88.9 \ \mu\text{C}$
Using the definition of capacitance, find the potential across the 10.0-μF capacitor.	$V_{10} = 8.89 \ \text{V}$
Using the definition of capacitance, find the potential across the 40.0-μF capacitor.	$V_{40} = 2.22 \ \text{V}$

Using the definition of capacitance, find the potential across the 100.0-μF capacitor.	
	$V_{100} = 0.889$ V

Check: The sum of the individual voltages should add up to 12.0 V, within the number of significant figures used.

Taking It Further: Explain why the largest capacitor has the smallest potential difference across its electrodes.

Try It Yourself #6

A 10.0-μF capacitor, a 40.0-μF capacitor, and a 100.0-μF capacitor are connected in parallel. A 12.0-V battery is connected across this combination. (a) What is the equivalent capacitance of the combination? (b) What is the charge on each capacitor? (c) What is the potential difference across each capacitor?

Picture: Use the expression for equivalent capacitance of capacitors in parallel. Capacitors in parallel all have the same potential difference, which can be used to determine the charge stored by each capacitor.

Solve:

Sketch the schematic of capacitors to visualize the circuit.	
Determine the equivalent capacitance for capacitors in parallel.	$C_{eq} = 150$ μF
The voltage across each capacitor and the battery are all the same since they are attached in parallel.	$V = 12.0$ V

Find the charge stored by the equivalent capacitance.	
Using the definition of capacitance, find the charge stored by the 10.0-μF capacitor.	$Q_{10} = 120 \ \mu$C
Using the definition of capacitance, find the charge stored by the 40.0-μF capacitor.	$Q_{40} = 480 \ \mu$C
Using the definition of capacitance, find the charge stored by the 100.0-μF capacitor.	$Q_{100} = 1.20$ mC

Check: The sum of the charges on the individual capacitors should add up to the total charge stored by the equivalent capacitance.

Taking It Further: What is the advantage to arranging several capacitors in parallel?

24.4   Dielectrics

In a Nutshell

A **dielectric** is simply an insulating material like air, glass, paper, wood, and others. When a region of space is filled with a dielectric other than vacuum the electrical properties of that space change. Mathematically, one change in electrical properties is described by the **dielectric constant** κ. Whenever an insulating material is present, we simply replace the permittivity of free space ϵ_0 with the permittivity of the dielectric material, which is given by $\kappa\epsilon_0$. All the physics about electrostatics we have discussed so far is valid in the presence of a dielectric material as long as we make this simple substitution.

Another advantage of dielectric materials is that they generally have a much larger dielectric strength than air, which means a capacitor filled with a dielectric other than air can withstand a higher potential difference across its electrodes than it could otherwise. Dielectrics are also useful for maintaining the separation between the electrodes of a capacitor, since they tend to attract each other.

The dielectric constant of empty space is 1. The dielectric constant of air is slightly larger than 1, but is so close that very little error occurs by using the value of 1 for it, as well. Table 24-1 on page 819 of the text lists the dielectric constant and dielectric strength of a variety of materials. The dielectric constant is always greater than or equal to 1.

The primary consequence of adding a dielectric between the electrodes of a capacitor is that the capacitance increases by a factor of κ over its air-filled value—that is, $C = \kappa C_0$. If a dielectric is inserted into an isolated capacitor, the electric field and electric potential between the electrodes are both reduced by a factor of κ.

Physical Quantities and Their Units

Dielectric constant $\kappa \geq 1$ and is dimensionless

Permittivity of a dielectric $\epsilon = \kappa\epsilon_0$

Important Derived Results

Capacitance with a dielectric $C = \kappa C_0$

Common Pitfalls

> ➤ Adding a dielectric between the plates of an isolated capacitor weakens the field in the capacitor gap because of the opposing electric field due to the bound charges on the surfaces of the polarized dielectric. This means that a given charge on the plates corresponds to a smaller potential difference, so the capacitance is increased.
> ➤ Mathematically, just remember to replace ϵ_0 with $\kappa\epsilon_0$ in the presence of a dielectric, and you should be fine.

7. TRUE or FALSE: An effect of inserting a dielectric between the plates of an isolated capacitor is to diminish the electric field of the capacitor.

8. Capacitors A and B are identical except that the region between the plates of capacitor A is filled with a dielectric. As shown in the figure, the plates of these capacitors are maintained at the same potential difference by a battery. Is the electric field intensity in the region between the plates of capacitor A smaller, the same, or larger than the field in the region between the plates of capacitor B? Explain.

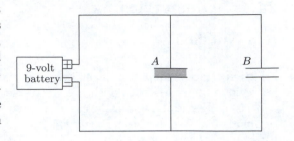

Try It Yourself #7

The plates of a parallel-plate capacitor are separated by a 0.500-mm-thick Pyrex sheet. The area of each plate is 3.00 m^2. (a) What is the maximum voltage between the plates before dielectric breakdown occurs? (b) What is the capacitance? (c) What is the maximum amount of electrostatic energy this capacitor can store?

Picture: The potential difference is the product of the electric field and the gap width. Use Table 24-1 on page 819 of the text to find the dielectric constant and dielectric strength of Pyrex. Dielectric breakdown occurs when the electric field exceeds the dielectric strength. The capacitance and potential energy stored can be found using the appropriate equations for a parallel-plate capacitor.

Solve:

Look up the dielectric strength of Pyrex in Table 24-1 on page 819 of the text.	
Relate the voltage on the plates to the dielectric strength of Pyrex and the separation of the plates. By setting the electric field to the dielectric strength of the material, you can find the maximum voltage.	$V_{\text{max}} = 7 \text{ kV}$
Use the expression for the capacitance of a parallel-plate capacitor filled with a dielectric to determine the capacitance.	$C = 0.297 \ \mu F$

Using the maximum voltage and capacitance found in the previous steps calculate the maximum electrical potential energy that can be stored by this capacitor.	
	$U_{\text{max}} = 7.29$ J

Check: The units all work out properly.

Taking It Further: What will happen if you succeed in pulling the Pyrex sheet out from between the plates?

Try It Yourself #8

Determine the capacitance of the plane parallel-plate capacitor shown. The dielectric with constant κ_1 fills up $\frac{1}{4}$ the area, but the full separation of the plates. The materials with constants κ_2 and κ_3 fill the other $\frac{3}{4}$ of the area, and divide the separation of the plates in half.

Picture: The region with κ_1 and κ_2 can be modeled as two capacitors in series. Their combination is in parallel with the region containing κ_3.

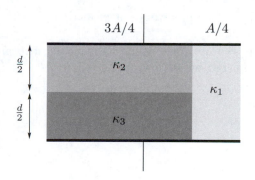

Solve:

Determine the capacitance of the region filled with dielectric constant κ_1.	
Determine the capacitance of the region filled with dielectric constant κ_2.	

Determine the capacitance of the region filled with dielectric constant κ_3.	
Find the equivalent capacitance of the combined space with κ_2 and κ_3. Treat them as if they are in series.	
Find the total equivalent capacitance. The region filled with κ_1 is in parallel with the capacitance calculated in the previous step.	$C_{\text{eq}} = \dfrac{\epsilon_0 A}{2d} \left(\dfrac{3\kappa_2 \kappa_3}{\kappa_2 + \kappa_3} + \dfrac{\kappa_1}{2} \right)$

Check: The units work out properly.

Taking It Further: What happens to the total capacitance if the material with dielectric constant κ_3 is replaced by air?

24.5 Molecular View of a Dielectric

In a Nutshell

If a dielectric is placed in an external electric field, the field pulls the positively and negatively charged particles (electrons and protons) within the dielectric in opposite directions. On average the electrons undergo a small displacement directed opposite to the field direction, and the positive nuclei undergo a small displacement in the direction of the field. These displacements result in surface charges on the dielectric called bound charges. They are so named because they are bound to individual atoms or molecules and are not free to move throughout the dielectric as are the free electrons in conductors.

In a parallel-plate capacitor with a dielectric occupying the space between the plates, the bound surface charge density σ_b results in an increase in capacitance by producing an electric field E_b within the dielectric that is smaller than, and oppositely directed to, the electric field E_0 produced by the free surface charge density σ_f on the surface of the metal plates.

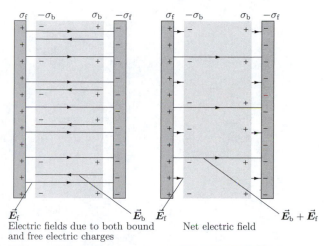

Electric fields due to both bound and free electric charges

Net electric field

In materials that exhibit the **piezoelectric effect** a mechanical stress on the material can also induce an electrical polarization of the material. This phenomenon can also be reversed—that is, applying a voltage across the material can cause it to expand or shrink.

Crystals that exhibit the **pyroelectric effect** have a large electric field generated within them when the crystal is heated.

Important Derived Results

Relation between bound and free surface charge densities $\qquad \sigma_b = \left(1 - \dfrac{1}{\kappa}\right)\sigma_f$

QUIZ

1. TRUE or FALSE: The capacitance of a charged capacitor is directly proportional to its charge.

2. TRUE or FALSE: The expression for electric energy density is $u_e = \frac{1}{2}\epsilon E^2$, where u_e is the energy per unit charge.

3. Why is the bound surface charge density on a dielectric always less than the free surface charge density on the capacitor plates?

4. Does inserting a dielectric into a capacitor increase or decrease the energy stored in the capacitor? Explain.

5. What are the benefits, if any, of filling a capacitor with a dielectric other than air?

6. You have a bucketful of capacitors, each with a capacitance of 1.00 μF and a maximum voltage rating of 250 V. You are to come up with a combination that has a capacitance of 0.75 μF and a maximum voltage rating of 1000 V. What is the minimum number of capacitors you need?

7. A parallel-plate capacitor has area A and separation d. How is its capacitance affected if a *conducting* slab of thickness $d' < d$ is inserted between, and parallel to, the plates as shown in the figure? Does your answer depend on where, vertically, between the plates the slab is positioned?

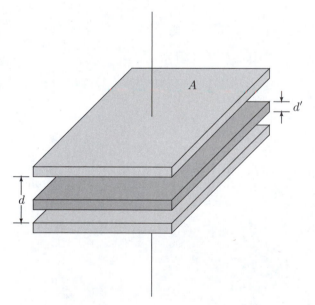

Chapter 25

Electric Current
and Direct-Current Circuits

25.1   Current and the Motion of Charges

In a Nutshell

Electric current is the time rate of flow of electric charge through a surface, such as the cross section of a wire. Thus if ΔQ is the charge flowing through a surface in time Δt, the current I through that area is $I = \lim_{\Delta t \to 0} \Delta Q/(\Delta t) = dQ/dt$. The SI unit of current is the **ampere** (A): 1 A = 1 C/s. By definition, the direction of positive current is the direction of motion of positive charges. This means that negatively charged electrons travel in the opposite direction to the current.

The flow of charge is caused by an electric field established by applying a potential difference. We generally think of this as occurring in wires, but it may also be used to generate a beam of electrons, alpha particles, or some other charged ion. In an electric circuit, a **direct current**, or **dc**, voltage can be established by batteries or some other dc power supply. In a dc circuit, once steady-state has been established the net direction of motion of the charged particles does not change.

The **drift velocity** is the net velocity of the moving charged particles. The **drift speed** is the magnitude of the drift velocity. The current in a wire is related to the drift speed according to $I = dQ/dt = qnAv_{\mathrm{d}}$, where A is the cross-sectional area of the wire, n is the **number density** of charge carriers (the number of mobile charged particles per unit volume), and v_{d} is the drift speed of the mobile charges.

Current is a scalar, even though it has a direction associated with it. A full vector representation of current is given by the **current density** $\vec{J} = nq\vec{v}_{\mathrm{d}}$. The **current** through a surface is defined as the flux of the current density vector through the surface, $I = \int_S \vec{J} \cdot d\vec{A}$.

Physical Quantities and Their Units

Current I SI units of ampere, 1 A = 1 C/s

Current density $\vec{J}$ SI units of A/m^2

Fundamental Equations

Current density $\vec{J} = nq\vec{v}_{\mathrm{d}}$

Current $I = \displaystyle\int_S \vec{J} \cdot d\vec{A} = \int_S \vec{J} \cdot \hat{n}\, dA = \dfrac{dq}{dt}$

Common Pitfalls

> ➤ Although the instantaneous speed of an electron in a metal is quite high ($\sim 10^6$ m/s), the drift speed is generally quite low—less than a mm/s.
> ➤ Because free charges are present throughout the volume of a metallic wire, charge begins to flow throughout the entire circuit almost immediately after a switch is closed.

1. TRUE or FALSE: In a metal the drift velocity of the free electrons is in the direction of the current.

2. We justified a number of electrostatic phenomena by the argument that there can be no electric field in a conductor. Now we say that the current in a conductor is driven by a potential difference and thus there is an electric field in the conductor. Is there a contradiction here?

Try It Yourself #1

A synchrotron radiation facility consists of a circular ring 40.0 m in radius and creates an electron beam with a current of 487 mA when the electrons have a speed approximately equal to the speed of light. How many electrons pass a given point in the accelerator in one hour?

Picture: Current is coulombs/second, which we are given. We need to convert this to electrons per hour.

Solve:

Write the current with units of coulombs per second.	
Determine how many electrons are required to achieve a magnitude of charge equal to 0.487 C. Your result should be in electrons per coulomb.	
Determine how many seconds are in each hour. Your result should be in seconds per hour.	

Being careful to watch your units, multiply everything together such that you end up with units of electrons per hour.	
	1.09×10^{22} electrons per hour

Check: That seems like a large number, but electrons don't carry much charge. The units all work out.

Taking It Further: Does this require that 1.09×10^{22} electrons are present in the synchrotron? Why or why not?

Try It Yourself #2

Current in a wire is given by the expression $I = (6.00 \text{ A}) \sin\left[\left(3.00 \text{ s}^{-1}\right) t\right] e^{-t/(2.00 \text{ s})}$. How much charge passes a point in the wire from $t = 0.500$ s to $t = 1.75$ s?

Picture: $I = dQ/dt$. Rearrange this expression and integrate to find the total charge.

Solve:

Write a generic, *algebraic* expression for dQ in terms of I and dt.	
Integrate both sides to find the total charge that passes. The integral of dQ is simply Q. Substitute the expression for I given in the problem. Don't forget your limits. You may want to review integration by parts.	
	$Q = 0.0599$ C

Check: The units work out properly.

Taking It Further: How many electrons is this?

25.2   Resistance and Ohm's Law

In a Nutshell

The internal electric fields that drive the current in a current-carrying metal wire are directed parallel to the current, opposite to the direction of electron flow. Because an electric field points in the direction of decreasing potential, the current is also in the direction of decreasing potential. The potential drop V—the decrease in potential along the direction of the current—across a length ΔL of the wire is $V = -\Delta V = \int \vec{E} \cdot d\vec{\ell} = E \, \Delta L$ because the electric field in a metal wire is approximately constant. This decrease in potential is exactly what a voltmeter hooked across this wire would measure.

The **resistance** R of a particular length of an object is defined as $R = V/I$, where V is the potential drop across the length and I is the current in the object. The SI unit of resistance, the volt per ampere, is called the **ohm** (Ω): $1\ \Omega = 1\ \text{V/A}$.

It is found that in metals the current is proportional to the potential drop across a given length of wire, that is, R is *constant*, independent of both V and I. This empirical result, which holds if the temperature of the metal remains fixed, is known as **Ohm's law**: $V = IR$, $R = \text{constant}$.

Empirically, the resistance of a length L of a conducting object is found to be proportional to the length of the object and inversely proportional to its cross-sectional area A: $R = \rho L/A$, where the proportionality constant ρ is the **resistivity**, measured in ohm-meters, of the conducting material. (The reciprocal of the resistivity is called the **conductivity** σ).

The resistivity of any given metal is not constant but varies with temperature. Empirically it is found that the resistivity changes fairly linearly with changes in temperature. The temperature coefficient of resistivity ρ, usually given at $T_0 = 20°C$, is the ratio of the change in resistivity to the change in temperature, so $\rho = \rho_0 \left[1 + \alpha \left(T - T_0\right)\right]$, where ρ is the resistivity at temperature T, ρ_0 is the resistivity at temperature T_0, and α is the **temperature coefficient of resistivity**. Table 25-1 on page 847 of the text lists the resistivities and temperature coefficients of several conducting materials.

A **resistor** is any electrical device specifically designed to cause a voltage drop along its length when charge flows through it.

Symbol for a resistor

The electric field and the current density in an object are related to that object's resistivity: $\vec{E} = \rho \vec{J}$.

Physical Quantities and Their Units

Resistance R SI units of ohms, $1\ \Omega = 1\ \text{V/A}$

Resistivity ρ SI units of $\Omega \cdot \text{m}$

Temperature coefficient of resistivity   SI units of $1/\text{K}$

Important Derived Results

Definition of resistance	$R = \dfrac{V}{I}$
Ohm's law	$V = IR,\ R = \text{constant}$
Resistance and resistivity	$R = \rho \dfrac{L}{A}$
Electric field and current density	$\vec{E} = \rho \vec{J}$
Temperature coefficient of resistivity	$\alpha = \dfrac{(\rho - \rho_0)/\rho_0}{T - T_0}$

Common Pitfalls

> ➤ Although *resistance* and *resistivity* sound similar, they are very different quantities. Resistance is a property of a specific object, but resistivity is a property of the material or materials the object is made of.
> ➤ Resistance depends both on geometry and on a material property, the resistivity. For example, the resistance of a wire with a uniform cross-sectional area A is $R = \rho L/A$.

3. TRUE or FALSE: For an ohmic material the resistance is independent of the current.

4. Two wires, A and B, have the same physical dimensions but are made of different materials. If A has twice the resistance of B, how do their resistivities compare?

Try It Yourself #3

A piece of 14-gauge copper wire 0.163 cm in diameter is 14.0 m long. (a) What is its resistance? (b) If a potential difference of 3.00 V is applied across the wire, what current flows in it? (c) What is the electric field in the wire? The resistivity of copper is $1.7 10^{-8}\ \Omega \cdot \text{m}$.

Picture: Calculate the resistance of the wire and use Ohm's law to find the current in the wire. The electric field in the wire is the voltage drop per unit length.

Solve:

Calculate the resistance of the wire. Make sure you convert your units properly.	
	$R = 0.114\ \Omega$

Use Ohm's law to find the current in the wire.	
	$I = 26.3$ A
Find the electric field.	
	$E = 0.214$ V/m

Check: It takes a very small field to produce a significant current in a wire. The units all work out properly.

Taking It Further: If the wire were made of silver instead of copper, how would your answers change?

25.3   Energy in Electric Circuits

In a Nutshell

In a current-carrying conductor the charge carriers lose potential energy but do not, on average, gain kinetic energy. Instead, they transfer energy to the metal. This dissipative process in which the potential energy of the charge carriers is transferred to the thermal energy of the conductor is called **Joule heating**. The rate at which thermal energy is generated equals the rate of loss of potential energy by the charge carriers. That is, $P = IV$. If this power is lost in a resistor, we can use the definition of resistance to find that $P = IV = I^2R = V^2/R$.

Because flowing charges continually lose kinetic energy, a constant supply of electrical energy is required. Devices that supply electrical energy to a circuit are called **sources of emf**. Common sources include batteries or generators. A source of emf does work on the charges passing through it, raising their electrical potential energy. The work per unit charge is called the emf $\mathcal{E}$ of the source. When a charge ΔQ flows through a source of emf, the work done on it is $+\mathcal{E} \, \Delta Q$. Thus the source of emf can be thought of as a "charge pump" raising the potential energy of the charge that flows through it. The unit of emf is the joule per coulomb, or volt. The power being delivered by a source of emf—that is, the rate at which electrical potential energy is generated, is $P = \mathcal{E}I$.

An **ideal battery** is a source of emf that maintains a constant potential difference between its two terminals, independent of the current through it. The potential difference between the terminals of an ideal battery is equal in magnitude to the emf. Such a device is also called a **voltage source**.

The potential difference across the terminals of a battery is called the **terminal voltage**. In an **ideal battery**, there are no internal losses of electrical energy, so the terminal voltage always equals the emf. In a **real battery**, the terminal voltage is not equal to the emf except when the current is zero. A real battery has an **internal resistance** r, which reduces the terminal voltage when current is present in the circuit. Thus if, as shown, a real battery is connected to an external resistor, the terminal voltage is $V_a - V_b = \mathcal{E} - Ir$, where a and b are, respectively, the $+$ and $-$ terminals of the battery. The potential drop across the external resistor is also $V_a - V_b$, so $V_a - V_b = IR$, where R is the **load resistance**, the resistance of the external resistor. So in this circuit the current is given by $I = \mathcal{E}/(R + r)$.

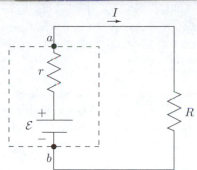

Diagram of battery and resistor pictured above

Batteries are typically rated by how much total charge they can provide to a circuit. The units typically used are ampere-hours $(\mathrm{A} \cdot \mathrm{h})$.

Important Derived Results

Rate of potential energy loss or gain	$P = IV$
Power dissipated by a resistor	$P = IV = I^2R = \dfrac{V^2}{R}$
Power supplied by an ideal emf source	$P = I\mathcal{E}$
Energy stored in a battery	$E_{\text{stored}} = Q\mathcal{E}$
Terminal voltage of a real battery	$V_a - V_b = \mathcal{E} - Ir$

Common Pitfalls

> When a battery is connected to a resistor, the current through the resistor is in the direction of the electric field within it, which means it is in the direction of decreasing potential. However, within the battery the electric field opposes the current, which means the current through it is in the direction of increasing potential.

> When using $P = IV = I^2R = V^2/R$ for determining the power dissipated in a resistor, be sure that I is the current *through* that resistor and V is the voltage drop *across* it in the direction of the current.

5. TRUE or FALSE: The equation $V = IR$ can be applied only to resistors made with ohmic materials.

6. A variable resistor is connected across the terminals of an ideal battery. Will the rate at which electrical energy is dissipated in the resistor increase or decrease as the resistor's resistance decreases? Why?

Try It Yourself #4

An electric heater consists of a single resistor connected across a 110-V line. It is used to heat 200.0 g of water in a cup (to make instant coffee) from 20°C to 90°C in 2.70 minutes. Assuming that 90 percent of the energy drawn from the power source goes into heating the water, what is the resistance of the heater?

Picture: The heat gained by the water is equal to 90 percent of the energy dissipated by the resistor. We can express the power dissipated by the resistor in terms of resistance and voltage and express the heat gained by the water in terms of its change in temperature.

Solve:

Find an *algebraic* expression for the power dissipated by the resistor in terms of the voltage and resistance.	
Find an *algebraic* expression for the total energy dissipated by the resistor, which is the power multiplied by the time over which that energy is dissipated.	
Find an expression for the heat gained by the water. You may want to refer to Chapter 18.	
Ninety percent of the energy dissipated by the resistor is equal to the heat gained by the water. Solve *algebraically* for the resistance of the heater.	

Substitute values and their units to find the resistance of the heater.	
	$R = 30.1\ \Omega$

Check: This is a reasonable value, and the units all work out.

Taking It Further: Assuming no other heat losses, how much longer will it take to heat your water if you have to power the water heater with your 12.0-V car battery?

Try It Yourself #5

An automotive battery has a terminal voltage of 12.5 V when it is delivering 30.0 A to the starter motor of a car. Under different load conditions it delivers 80.0 A and its terminal voltage is 10.7 V. Find the constant internal resistance and the emf of the battery.

Picture: In each case, the circuit diagram consists of the emf of the battery, the internal resistance of the battery, and the resistance of the external load. The terminal voltage of the battery is equal to its emf minus the voltage drop across its internal resistance.

Solve:

Draw a diagram of the circuit. Let $\mathcal{E}$ be the emf of the battery and r its internal resistance.	
Find an *algebraic* expression for the terminal voltage.	
Substitute into the above expression values with units for the two cases described in the problem.	

Subtract the two expressions obtained in the previous step to find the internal resistance r.	$r = 0.0360 \ \Omega$
Substitute the value for the internal resistance into one of the expressions obtained in the third step to find the emf of the battery.	$\mathcal{E} = 13.6 \ \text{V}$

Check: Both these values seem reasonable, and the units work out.

Taking It Further: What is the value of the load resistance in each case? Why does the 80.0 A case have a smaller load resistance?

25.4   Combinations of Resistors

In a Nutshell

Two or more resistors are said to be connected **in series** when they are connected as shown. Each resistor connected in series in a circuit carries the same current, and the potential drop across a combination of resistors connected in series is the sum of the potential drops across each of the individual resistors in the combination. The equivalent resistance of a combination of resistors is the resistance of a single resistor that would give the same potential drop V as the combination when carrying the same current I, so the equivalent resistance is $R_{\text{eq,series}} = R_1 + R_2 + R_3 + \ldots$.

Two or more resistors are said to be connected **in parallel** when they are connected as shown. Each resistor connected in parallel has the same potential drop across it, and the total current through the combination equals the sum of the currents through each of the individual resistors in the combination; so the equivalent resistance is $1/R_{\text{eq},\parallel} = (1/R_1) + (1/R_2) + (1/R_3) + \dots$.

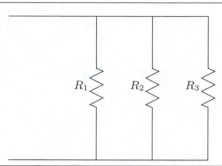

Problems Involving Series and/or Parallel Combinations of Resistors

Picture: If no circuit diagram is provided, draw one.

Solve:

1. Identify each series and/or parallel combination of resistors and calculate its equivalent resistance.
2. Redraw the circuit so that each series or parallel combination of resistors is replaced by a single resistor of equivalent resistance.
3. Repeat steps 2 and 3 until there are no more series or parallel combinations. (At this point the circuit should contain only a single resistor.) Apply $V = IR$ and calculate the current.
4. Return to the previous drawing and calculate the voltage across and/or the current in each resistor in that drawing.
5. Repeat step 4 until you have calculated all currents and/or voltages of interest.

Check: Calculate the power delivered to each resistor using $P = IV$ or its equivalent and calculate the power supplied by the sources of emf using $P = I\mathcal{E}$. Then check to see that the total power delivered equals the total power supplied.

Important Derived Results

Equivalent series resistance $\qquad\qquad\qquad R_{\text{eq,series}} = R_1 + R_2 + R_3 + \dots = \sum_i R_i$

Equivalent parallel resistance $\qquad\qquad \dfrac{1}{R_{\text{eq},\parallel}} = \dfrac{1}{R_1} + \dfrac{1}{R_2} + \dfrac{1}{R_3} + \dots = \sum_i \dfrac{1}{R_i}$

Common Pitfalls

> Remember that the rules for the equivalent resistance of resistors in series and parallel are exactly opposite to the rules for combining capacitors.

> The rules for the equivalent resistance of series and parallel combinations of resistors apply only for those exact combinations. It's easy to look at two resistors in a circuit, for instance, and think they're in series when actually another branch forks off between them.

> When you calculate the equivalent resistance of a parallel combination by adding the inverse resistances, don't forget that you have to invert the result to obtain the equivalent. Units can help here.

7. TRUE or FALSE: In a metallic resistor the free electrons move from low potential to high potential.

8. Many ordinary strings of Christmas-tree lights contain 50 or so bulbs connected in parallel across a 110-V line. Forty years ago most strings contained 50 bulbs connected in series across the line. What would happen if you could put one of the old-style bulbs into a modern Christmas-tree light set? (The light sockets are made differently to prevent this.)

Try It Yourself #6

In the circuit shown a potential difference of 5.00 V is applied between points a and b. Find (a) the equivalent total resistance, (b) the current in each resistor, and (c) the power being dissipated in each resistor.

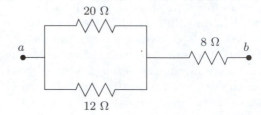

Picture: You will need to find two equivalent resistances, first for the two resistors in parallel and then one for all three resistors. Remember that the voltage across resistors in parallel is the same for all resistors, and the current is the same through all resistors in series.

Solve:

Find the equivalent resistance of the two resistors in parallel.	
The equivalent resistance calculated in the first step is in series with the 8.00 Ω resistor. Calculate the total equivalent resistance of this series combination.	$R_{\text{eq}} = 15.5 \ \Omega$
Using Ohm's law, find the total current in the equivalent resistance. This total current is the same current that flows through the 8.00-Ω resistor.	$I_{\text{total and 8 }\Omega} = 0.323$ A
Find the potential difference across the two resistors in parallel, using their equivalent resistance calculated in the first step and the total current just calculated.	

Use this potential difference to determine the current in each of the two parallel resistors.	$I_{20\,\Omega} = 0.121$ A, $I_{12\,\Omega} = 0.202$ A
Knowing the resistances and the current in each resistor, you can calculate the power dissipated in each resistor.	$P_{20\,\Omega} = 0.293$ W, $P_{12\,\Omega} = 0.490$ W, $P_{8\,\Omega} = 0.832$ W

Check: The units all check out.

Taking It Further: Why is the least amount of power dissipated in the largest resistors?

Try It Yourself #7

A potential difference of 7.50 V is applied between points a and c in the circuit shown. Find the difference in potential between points b and c.

Picture: The given voltage is across the combination of the 35-Ω and 65-Ω resistors, which are in series. Once you know the current through the equivalent series resistance, you can use Ohm's law a second time to find the voltage drop across the 65-Ω resistor alone.

Solve:

Find the equivalent resistance of the two resistors in series.	

Use Ohm's law to determine the current through the equivalent resistance required to maintain the given potential difference.	
The current through the equivalent resistance is the same as the current through each resistor in series. Use the current from step 2 to find the potential difference across the 65-Ω resistor.	$V = 4.88$ V

Check: The potential difference is more than half the total potential difference. This is expected because 65 Ω is more than half the total equivalent series resistance.

Taking It Further: Is the current through the 60-Ω resistor larger or smaller than that through the 35-Ω resistor? Why?

25.5   Kirchhoff's Rules

In a Nutshell

Many circuits, such as the one shown, cannot be simply reduced to series and parallel resistors. For these circuits more robust analysis techniques are required. One such technique is known as the application of Kirchhoff's rules.

Kirchhoff's first rule states, "When any closed loop is traversed, the algebraic sum of the changes in potential around the loop must equal zero." This rule, also known as the **loop rule**, was introduced in Chapter 24. It is a direct result of conservation of energy. The electric field responsible for the flow of charge is a conservative force. So any charge that starts and ends at the same location must experience no net change in potential energy. This means there must also be no net change in electrical potential V.

Kirchhoff's second rule states, "At any junction (branch point) in a circuit where the current can divide, the sum of the currents into the junction must equal the sum of the currents out of the junction." This rule, also known as the **junction rule**, follows directly from the conservation of charge. If the second rule were not true, then charge could accumulate arbitrarily at junctions. This would eventually result in no current flow, which simply does not happen.

When applying Kirchhoff's rules you must assign a positive direction for the current in each segment of the circuit. By definition you do not necessarily know the direction of the current in each segment of the circuit a priori. This is fine. Pick a direction and label for each current and stick with it. If you guessed the wrong direction you will simply end up with a negative value for that current.

Several sign conventions are required when applying Kirchhoff's rules to a circuit. The first of these is the following: When traversing a source of emf from its negative terminal to its positive terminal, the change in potential is positive and equal to the emf of the source. When traversing a source of emf from its positive terminal to it negative terminal, the change in potential is negative, and equal in magnitude to the emf of the source.

The sign convention for the change in potential across a resistor is slightly more complicated. If you traverse a resistor in the direction of the positive current you chose above, the change in potential $\Delta V = -IR$. If you traverse a resistor in the direction opposite to the direction of positive current then the change in potential $\Delta V = +IR$.

Method for Analyzing Multiloop Circuits

Picture: Draw a diagram of the circuit.

Solve:

1. Replace any series or parallel resistor combinations or capacitor combinations with their equivalent values.
2. Repeat step 1 as many times as possible.
3. Next, assign a positive direction for each branch of the circuit and indicate this direction with an arrow. Label the current in each branch. Add plus and minus signs to indicate the high-potential and low-potential terminals of each source of emf.
4. Apply the junction rule to all but one of the junctions.
5. Apply the loop rule to the different loops until you obtain as many independent equations as there are unknowns. When traversing a resistor in the positive direction, the change in potential equals $-IR$. When traversing a battery from the negative terminal to the positive terminal, the change in potential equals $\mathcal{E}Ir$.
6. Solve the system of equations to obtain the desired values.

Check: Check your results by assigning a potential of zero to one point in the circuit and use the values of the currents found to determine the potentials at other points in the circuit.

An **ammeter** is used to measure current. It is placed "inline" or in series with other circuit elements. An ideal ammeter has zero resistance so that it has zero impact on the circuit it is measuring. Real ammeters typically have internal resistances of less than an ohm. An ammeter is represented by a circle with a letter "A" in it.

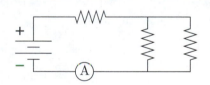

A **voltmeter** is used to measure potential differences in a circuit. A voltmeter is attached "across" or parallel to the circuit element(s) whose potential difference you wish to measure. An ideal voltmeter has an infinite resistance so that it has zero impact on the circuit it is measuring. Real voltmeters typically have an internal resistance of 100 MΩ or more. A voltmeter is represented by a circle with a letter "V" in it.

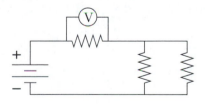

An **ohmmeter** is used to measure the resistance of an object. An ohmmeter works by applying a known voltage to the object, measuring the resulting current, and applying $V = IR$ to the result to determine the resistance. Because this measurement relies on applying a known voltage, resistors must be removed from the circuit in order to get an accurate measurement.

Important Derived Results

Kirchhoff's loop rule

$$\sum_{\text{closed loop}} \Delta V = 0$$

Kirchhoff's junction rule

$$\sum I_{\text{in}} = \sum I_{\text{out}}$$

Common Pitfalls

> When you apply Kirchhoff's loop rule, remember that the potential decreases as you go past a resistor in the direction of the current but increases if you are going "upstream." Likewise, the voltage increases as you go through a source of emf from the negative to the positive terminal, and vice versa. It's easy to make sign errors. It will help to ask yourself whether or not you are moving in the direction of the electric field. The electric field always points in the direction of decreasing potential.

> You have to assign a direction to the current in each branch of a circuit to apply Kirchhoff's rules. It doesn't matter at all if the actual direction of one or more currents is opposite to the direction you assign it. When you solve the problem, that current will come out negative.

> A voltmeter measures the potential difference between its terminals. Its terminals are connected to the two points between which you want to know the potential difference. An ammeter measures the current through it. It must be connected so that the current you want to measure flows through the meter.

9. TRUE or FALSE: An ideal ammeter has a very large internal resistance.

10. On a hot day a co-worker needed to measure the voltage at a wall outlet. He set his multimeter (a combination voltmeter, ohmmeter, and ammeter) to measure voltage and connected its terminals to a wall outlet. The voltage reading was about 112 V. He then asked, "I wonder what the current is?" and changed the setting of the meter to measure current. What do you think happened when the meter became an ammeter? Don't try this at home.

Try It Yourself #8

For the circuit shown, find (a) the current in each resistor, (b) the power supplied by each source of emf, and (c) the power dissipated in each resistor.

Picture: There are three unknown currents, so we need three equations. Therefore, apply the junction rule to one of the junctions, and apply the loop rule to the two interior loops to solve for the current in each segment of the circuit. Once you know the current, you can calculate the power dissipated by each resistor and provided by each source of emf.

Solve:

Draw the circuit again, with labels for the current in each branch and arrows indicating your guess for the direction of current flow in each branch. Let I_1 be positive to the right through the 1 Ω resistor, I_2 be positive upward through the 2 Ω resistor, and I_3 be positive downward through the 10 Ω resistor. Place plus and minus signs indicating the polarity for each resistor based on your directions for the current flow.	
Apply the junction rule to the top junction.	
Apply the loop rule to the leftmost loop.	
Apply the loop rule to the rightmost loop. Reduce this expression by dividing by 2.	

Solve the equation in the second step for I_1.	
Substitute the expression from the previous step into the equation from step 3 and simplify. Multiply the entire equation by 2 to assist in the next step.	
Multiply the result for the rightmost loop by 7 and add it to the result from the previous step, solving for I_3.	
Substitute this value back into the result for the rightmost loop to solve for I_2.	
Substitute values for I_2 and I_3 into the junction rule to solve for I_1.	
Find the power supplied by each source of emf.	$P_{6\ \mathrm{V}} = 8.84$ W, $P_{2\ \mathrm{V}} = 2.95$ W, $P_{4\ \mathrm{V}} = -3.58$ W

Determine the power dissipated by each resistor.	
	$P_{1\ \Omega} = 2.17$ W, $P_{0.5\ \Omega} = 1.09$ W, $P_{2\ \Omega} = 1.60$ W, $P_{10\ \Omega} = 3.35$ W

Check: The power dissipated by the resistors equals the power provided by the batteries, so energy is conserved.

Taking It Further: What is the meaning of negative power for the 4-V battery?

Try It Yourself #9

Find the current in each of the three resistors of the circuit shown.

Picture: Use the same steps you used to find the currents in the previous example.

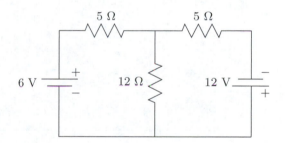

Solve:

Draw the circuit again, with labels for the current in each branch, and arrows indicating your guess for the direction of current flow in each branch. Place plus and minus signs indicating the polarity for each resistor based on your directions for the current flow.	
Apply the junction rule to the top junction.	

Apply the loop rule to the leftmost loop.	
Apply the loop rule to the rightmost loop.	
Solve the three equations for the three unknown currents.	$I_{5\ \Omega,\text{left}} = 1.70$ A, to the right; $I_{5\ \Omega,\text{right}} = 1.90$ A, to the right; $I_{12\ \Omega} = 0.207$ A, upward

Check: The units work out, and it is easy to confirm that the junction rule holds with these values.

Taking It Further: Current is always a positive quantity. What, then, is the meaning of a negative current when solving Kirchhoff's equations?

25.6   *RC* Circuits

In a Nutshell

An ***RC* circuit** is a circuit that contains both a resistor and a capacitor. The current in *RC* circuits is in a single direction, as with all dc circuits, but the magnitude of the current changes with time.

To apply Kirchhoff's loop rule to a circuit that contains a capacitor, we first arbitrarily label the charge on one plate of the capacitor $+Q$; the other plate, of course, has charge $-Q$. If it is given that a particular plate is positively charged, then it seems appropriate to label that charge $+Q$. If the sign of the charge is not given, a judicious choice is to label as $+Q$ the charge on the plate that becomes more positively charged when the current I enters it, because the current and the charge are then related as shown in Figure A. Otherwise, the charge Q and the current I are related as shown in Figure B.

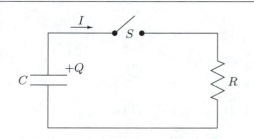

(A)

(B)

After the switch is closed in the circuit shown in the figure, the charge on the discharging capacitor Q is related to the resistance, the capacitance, and the elapsed time according to $Q(t) = Q_0\,e^{-t/RC}$. The quantity $\tau = RC$ is called the **time constant** of the circuit; it is the time it takes for the charge to decrease to $1/e$ of its initial value Q_0. The current in the circuit can be found from the negative time derivative of the charge on the capacitor. The negative sign is required because the charge Q on a discharging capacitor is being reduced, which means $dQ/dt < 0$, but current must be positive. The result is $I = I_0\,e^{-t/\tau}$, where $I_0 = V_0/R$, and V_0 is the initial voltage across the capacitor.

The capacitor in this figure is initially uncharged. After the switch is closed, the current charges the capacitor. By applying Kirchhoff's loop rule and solving for the charge, we have $Q(t) = C\mathcal{E}(1 - e^{-t/RC}) = Q_{\mathrm{f}}(1 - e^{-t/\tau})$, where Q_{f} is the final charge as $t \to \infty$. For this circuit I is the rate of increase of the charge Q, so $I = +dQ/dt = (\mathcal{E}/R)\,e^{-t/\tau} = I_0\,e^{-t/\tau}$, where $I_0 = \mathcal{E}/R$ is the current immediately after the switch is closed at $t = 0$.

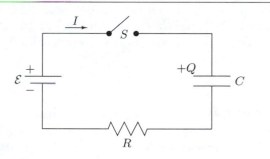

Important Derived Results

Capacitor discharging through a resistor

$$Q(t) = Q_0\,e^{-t/RC} = Q_0\,e^{-t/\tau}$$

Capacitor charging through a resistor

$$Q(t) = C\mathcal{E}\left(1 - e^{-t/RC}\right) = Q_{\mathrm{f}}\left(1 - e^{-t/\tau}\right)$$

Common Pitfalls

> When a capacitor is charged through a resistor, not all the potential energy delivered by the battery goes into charging the capacitor; some of it is dissipated in the resistor.

> If the actual polarity of a capacitor is opposite to the polarity that you assign, when you solve for the charge (or voltage) it will come out negative.
> To choose the correct sign in the relation $I = dQ/dt$ you must consider whether a particular current I increases or decreases the charge Q on a capacitor.
> When a capacitor becomes fully charged it acts like an open switch. No current can exist in the circuit path containing the capacitor.
> An uncharged capacitor acts like a short circuit or perfect wire. There is no voltage drop across it.
> Although the charge on the capacitor has a different time dependence for the charging and discharging cases, the current in an RC circuit has the same functional form in both cases.

11. TRUE or FALSE: Kirchhoff's rules apply to circuits that contain only ohmic materials.

12. Give a simple physical explanation why the charge on a capacitor in an RC circuit can't be changed instantaneously.

Try It Yourself #10

For the circuit shown, find (a) the current through the battery just after the switch is closed, (b) the steady-state current through the battery after the switch has been closed a long time, and (c) the maximum charge on the capacitor.

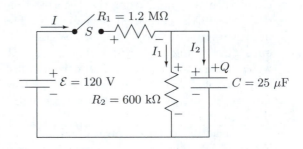

Picture: Initially, all the current goes into the capacitor and none through R_2 because there will be no charge and hence no potential difference across the capacitor or R_2. After a long time, the capacitor will be fully charged, so no more charge will flow into it.

Solve:

Initially, the charge on the capacitor is zero, so the voltage across its plates is zero. Apply Kirchhoff's loop rule to the outside loop to find the initial current I. $I = 100\ \mu A$	
In the steady state, no current flows through the capacitor, which means that $I = I_1$. Apply Kirchhoff's loop rule to the left hand loop to find the steady-state current provided by the battery. $I = 66.7\ \mu A$	

Determine the potential difference across R_2 in the steady-state condition. This is equal to the voltage across the capacitor when it is fully charged.	
	$V_C = V_{R_2} = 40.0$ V
Use the final voltage across the capacitor to determine the final charge on the capacitor.	
	$Q_f = 1.00$ mC

Check: All the units work out properly.

Taking It Further: Draw the circuit again, showing how you would place an ammeter to measure the current I and a voltmeter to measure the potential difference across R_1.

Try It Yourself #11

For the circuit shown, just after the switch is closed find (a) the current through the battery and (b) the current through the 200-kΩ resistor. (c) Find the steady-state charge on the capacitor after the switch has been closed a long time.

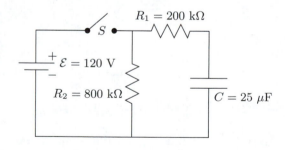

Picture: This problem is very similar to the previous example.

Solve:

Make a diagram of the circuit, labeling the current in each segment, and indicating the direction of current flow. Label the polarity of the capacitor.	

Initially, the charge on the capacitor is zero, so the voltage across its plates is zero. Apply Kirchhoff's loop rule to the outside loop to find the initial current through the capacitor and R_1.	$I_{R_1} = 0.600$ mA, to the right
Apply Kirchhoff's loop rule to the leftmost loop to find the initial current through R_2.	
Use Kirchhoff's junction rule to find the total current provided by the battery.	$I_{\text{total}} = 0.750$ mA, upward
In the steady state, no current flows through the capacitor, so no current will flow through R_1, either. Apply Kirchhoff's loop rule to the outside loop to find the final charge on the capacitor.	$Q_{\text{final}} = 3.00$ mC

Check: All the units work out properly.

Taking It Further: Draw the circuit again, showing how an ammeter would be placed to measure the current through R_2 and the voltage across C.

QUIZ

1. TRUE or FALSE: The drift velocity of the free electrons in a current-carrying wire refers to the average speed of the free electrons.

2. TRUE or FALSE: The time constant associated with the discharge of a capacitor through a resistor is R/C.

3. Does the time required to charge a capacitor through a given resistor with a battery depend on the emf? Does it depend on the total amount of charge to be placed on the capacitor?

4. The average (drift) velocity of electrons in a wire carrying a steady current is constant even though the electric field within the wire is doing work on the electrons. What happens to this energy?

5. When 120 V is applied to the filament of a 75-W light bulb, the current drawn is 0.63 A. When a potential difference of 3 V is applied to the same filament, the current is 0.086 A. Is the filament made of an ohmic material? Explain.

6. For a silver wire 0.100 inch in diameter and 100 feet long, carrying a current of 25.0A, find (a) the resistance, (b) the potential difference between the ends of the wire, (c) the electric field in it, and (d) the rate at which heat is generated in the wire.

7. A 12-V battery with an internal resistance of 0.600 Ω is used to charge a 0.200-μF capacitor through a 5.00-Ω resistor. Find (a) the initial current drawn from the battery, (b) the time constant of the circuit, and (c) the time required to charge the battery to 99% of the final charge.

Chapter 26

The Magnetic Field

26.1   The Force Exerted by a Magnetic Field

In a Nutshell

Magnetism is a fundamental property of matter. Naturally occurring magnetic forces have been known for over 2000 years. Every magnet has two regions, called poles, where the magnetic field is strongest. The poles are designated "north" and "south" since a freely suspended magnet orients itself in approximately a north–south direction because Earth itself has a magnetic core. The north poles of magnets point toward the north geographic pole of Earth, which is actually a magnetic south pole. Like poles of two magnets repel each other and opposite poles attract, with a force that is inversely proportional to the square of the distance between them. Magnetic poles exist only in equal and opposite pairs; and the observation of a single isolated pole, a monopole, has never been confirmed. In this chapter we focus on the effects of a given magnetic field. In Chapter 27 we will see that the source of the magnetic field is really an electric current.

A compass needle is an arrow-shaped magnet with its tail as the south pole and its tip as the north pole. The existence of a magnetic field $\vec{B}$ at any point in space can be demonstrated by suspending a compass needle there. The magnetic field will exert a torque on the needle, aligning it so that its north end points in the direction of the field.

Experimentally we observe that a magnetic field exerts a force on a charged particle only if the particle is in motion. The force is proportional to the charge of the particle and the strength of the magnetic field. Furthermore, the magnetic force is *perpendicular* to both the velocity $\vec{v}$ of the particle and the magnetic field $\vec{B}$. These empirical observations lead us to describe the magnetic force $\vec{F}$ on a particle of charge q moving with velocity $\vec{v}$ in a region with magnetic field $\vec{B}$ using a cross product relationship: $\vec{F} = q\vec{v} \times \vec{B}$. This equation defines the magnetic field in terms of the force exerted on a moving charge. The force is in the direction of the cross product for positively charged particles and opposite to the direction of the cross product for negatively charged particles. The direction of the cross product can be obtained by the right-hand rule, as shown.

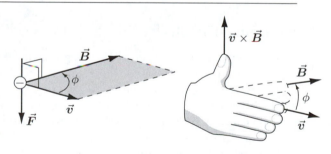

The SI unit of the magnetic field is the tesla (T). $1\text{ T} = 1\text{ N}/(\text{C} \cdot \text{m/s}) = 1\text{ N}/(\text{A} \cdot \text{m})$. Another commonly used unit of magnetic field is the gauss (G), where $1\text{ G} = 10^{-4}\text{ T}$. The magnetic field strength of Earth is slightly less than 1 G. Magnetic fields are often given in gauss because the unit of tesla is so large.

The magnetic force on a current-carrying wire in a magnetic field is the sum of the forces on all the moving charge carriers in the wire. The net magnetic force on a very small segment of current-carrying wire of length $d\vec{\ell}$ is $d\vec{F} = I\,d\vec{\ell} \times \vec{B}$, where the vector $d\vec{\ell}$ is parallel to the direction of the current ($\vec{J}$). The quantity $I\,d\vec{\ell}$ is called a **current element**.

Just as the electric field $\vec{E}$ can be represented by electric field lines, the magnetic field $\vec{B}$ can be represented by magnetic field lines. However, whereas electric field lines originate on positive charges and terminate on negative charges, magnetic field lines have no beginnings or ends. They are continuous and wrap back around onto themselves. The figure illustrates the differences between electric and magnetic fields of electric (left) and magnetic (right) dipoles.

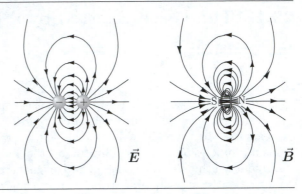

Physical Quantities and Their Units

Magnetic field   SI units of tesla (T)

Fundamental Equations

Magnetic force on a charged particle $\vec{F} = q\vec{v} \times \vec{B}$

Magnetic force on a current element $d\vec{F} = I\,d\vec{\ell} \times \vec{B}$

Common Pitfalls

> The behaviors of electric and magnetic fields are fundamentally different. The electric field is always parallel to—and the magnetic field always perpendicular to—the direction of the force exerted on a charged particle.

> What we call the north pole of a magnet is more appropriately called a north-seeking pole because it seeks the geographic north pole of Earth when the magnet is suspended. Because opposite poles attract and like poles repel, it is a south magnetic pole near the geographic north pole that attracts the north pole of the compass.

> Remember that the tangent to the magnetic field lines points in the direction of the magnetic field $\vec{B}$, not in the direction of the magnetic force $\vec{F}$ exerted on a moving electric charge.

1. TRUE or FALSE: If the direction of a charged particle's velocity is the same as the direction of the magnetic field, the magnetic force does work on the particle.

2. A current-carrying wire is in a region where there is a magnetic field, but there is no magnetic force acting on the wire. How can this be?

Try It Yourself #1

A horizontal, current-carrying wire of mass 40.0 g is free to slide without friction along two vertical conducting rails spaced 80.0 cm apart as in the figure. A uniform magnetic field of 1.20 T is directed into the plane of the drawing. What magnitude and direction must the current have if the force exerted by the magnetic field on the wire is just sufficient to balance the force of gravity on the wire?

Picture: The net force on the wire is zero. Use the expression for the magnetic force on a current element.

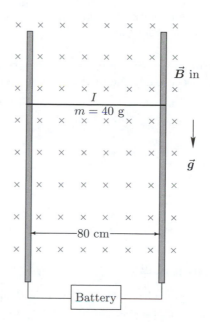

Solve:

Draw a free-body diagram of the wire. For the net force on the wire to be zero, the magnetic force must be directed upward. Use the right-hand-rule to determine the direction of current flow that results in an upward magnetic force.	
	Current in the bar must be to the right.
Write an *algebraic* expression for the magnetic force on the bar	
Set up the vector form of Newton's second law and solve *algebraically* for the current in the wire.	
Substitute values with units into your expression above and find the magnitude of the required current.	
	$I = 0.409$ A

Check: The units work out and this is a reasonable current. The right-hand rule gives the proper direction for the magnetic force.

Taking It Further: If the magnetic field were directed out of the page, how would your answers differ? What if the magnetic field were directed upward?

Try It Yourself #2

A segment of wire with total length L is situated in the second quadrant of the x,y plane, as shown, making an angle θ with respect to the $-x$ axis, and carrying a current I in the direction shown. It lies in a *non-uniform* magnetic field $\vec{B} = -(B_0/L)x\hat{k}$, which is in the positive z direction since $x < 0$ for the wire. What is the magnetic force on this line of current?

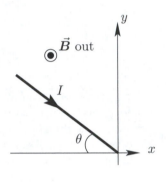

Picture: Use the expression for the differential force on a current element. As long as you are careful with the limits of integration, you can always write $d\vec{\ell} = dx\,\hat{i} + dy\,\hat{j}$.

Solve:

Write a generic, *algebraic* expression for the differential force on a current element.	
Find an expression for the y location of a current element as a function of x and θ. Be careful with your signs. Remember that x is a coordinate.	
Differentiate the above expression and write the result here, explicitly including the "$dy/dx =$" part of the expression. Then multiply both sides by dx to arrive at an expression for dy in terms of dx.	

Substitute this expression for dy into the expression for $d\vec{\ell}$ given in the Picture stage above.	
Substitute your expression for $d\vec{\ell}$ from the previous step and the given magnetic field into the expression for the differential force on a current element. Then perform the cross product, being careful to watch your signs.	
Integrate the expression obtained above. Be careful with your limits of integration over x. You should integrate x in the direction of the current. You actually have two integrals to perform—one for the $\hat{\imath}$ direction and one for the $\hat{\jmath}$ direction.	$$\vec{F} = \left[-\frac{1}{2} I B_0 L \cos(\theta) \sin(\theta) \right] \hat{\imath}$$ $$- \left[\frac{1}{2} I B_0 L \cos^2(\theta) \right] \hat{\jmath}$$

Check: The right-hand rule tells us that the magnetic force should be in the $-x$ and $-y$ directions, and our answer exhibits that behavior.

Taking It Further: How would your approach to this problem change if $\vec{B}$ were a function of y instead of x?

26.2   Motion of a Point Charge in a Magnetic Field

In a Nutshell

The magnetic force on a moving charged particle always acts at right angles to the particle's velocity. Thus the magnetic force does no work on the particle and has no effect on the particle's kinetic energy. A magnetic force can change the direction of the velocity of a particle but *not* its magnitude.

Consider an otherwise free charged particle moving with its velocity perpendicular to a uniform magnetic field. The magnetic force has a constant magnitude and acts at right angles to the velocity of the particle. In accordance with Newton's second law ($\vec{a} = \sum \vec{F}/m$), such a particle moves in uniform circular motion. Applying Newton's second law to such a particle results in the expression $qvB = mv^2/r$, where q is the charge, v the speed, B the magnetic field, and r the radius of the circle. The circumference of the circle equals the speed times the time for one revolution (the period T)—that is, $2\pi r = vT$. The frequency f of the motion is the reciprocal of the period, $f = qB/(2\pi m)$. Note that by taking advantage of Newton's second law we can see that the period and frequency do not depend on either the speed of the particle or the radius of the orbit. Exploiting this result, in the early 1930s E. O. Lawrence and M. S. Livingston built a particle accelerator, the earliest cyclotron, in which the particles move in two semicircular structures called dees (because they are in the shape of the letter "D"). Consequently the frequency and period of particles moving in an orthogonal magnetic field are called the **cyclotron frequency** and **cyclotron period**.

If the velocity of a charged particle is not perpendicular to the magnetic field, the particle moves in a helical, corkscrewlike, path. We can resolve the velocity into a component $v_\parallel$ that is parallel to the field and a component $v_\perp$ perpendicular to the field. The motion due to $v_\perp$ is the same as that just discussed. In accordance with the equation $\vec{F} = q\vec{v} \times \vec{B}$, the magnetic force acts perpendicular to $\vec{v}$, so $v_\parallel$ is unaffected by the magnetic field.

Many experiments require a beam of charged particles in which each particle in the beam moves at the same velocity. A **velocity selector** is a device that can select charged particles moving at a specific velocity while deflecting those moving either faster or slower. It consists of a region of space that contains a mutually perpendicular electric field $\vec{E}$ and magnetic field $\vec{B}$. As shown in the figure, a charged particle enters this region with a velocity perpendicular to both $\vec{E}$ and $\vec{B}$. The net force on this particle is zero only when the electric and magnetic forces are equal in magnitude and oppositely directed. Equating the magnitudes of these forces, we have $v = E/B$. Charged particles with this speed pass through the region undeflected. Particles moving either more quickly or more slowly are deflected because the magnetic force is not equal in magnitude to the electric force.

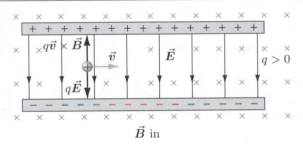

A **mass spectrometer** is used to measure the masses of charged particles such as molecules and ions. In a mass spectrometer a particle with known charge q moving with known speed v enters a region of uniform magnetic field $\vec{B}$, where it moves in uniform circular motion. The mass of the particle determines the radius r of its circular trajectory.

Important Derived Results

Cyclotron period $$T = \frac{2\pi m}{qB}$$

Cyclotron frequency $$f = \frac{1}{T} = \frac{qB}{2\pi m}$$

Common Pitfalls

> The magnetic force acts only on charges that have a component of velocity perpendicular to the magnetic field. A charge at rest or moving along the field direction experiences no magnetic force.

> The general motion of a charged particle in a uniform magnetic field is in a helical path with the axis of the helix in the field direction. In the special case in which the velocity of a particle is directed perpendicularly to the field direction, the motion is uniform circular motion. The period of this motion depends on the magnetic field strength and the charge and mass of the particle. It does *not* depend on the speed of the particle or the radius of the circular trajectory; if the speed varies, the radius changes proportionally.

3. TRUE or FALSE: In a cyclotron, as the speed of the particle increases the ratio of the speed to the circumference of the orbit remains constant.

4. For the velocity selector shown above, describe the motion of an electron if it travels (a) faster than E/B and (b) more slowly than E/B.

Try It Yourself #3

A cyclotron used to accelerate protons has a uniform magnetic field of 1.10 T. (a) What is the frequency at which the potential difference between the dees must oscillate? (b) If the inner radius of the dees is 30.0 cm what is the maximum kinetic energy (in MeV) attainable by the protons? (c) If the protons gain 72.0 keV of kinetic energy per revolution, what is the potential difference between the dees as the protons transit from one dee to the other?

Picture: Refer to pages 898–899 of the text to refresh your memory regarding the basic operating principles of a cyclotron. The oscillation frequency of the potential difference between the dees should be the same as the cyclotron frequency. Use Newton's second law to relate the radius of the orbit with the speed, and hence kinetic energy, of the protons. Remember that the protons are sped up each time they cross the gap between the dees, which occurs twice per revolution.

Solve:

Calculate the cyclotron frequency, which is the same as the frequency of the potential difference between the dees. Look up the mass of the proton in the table of physical constants in the back of the text.	
	$f = 16.8$ MHz

Apply Newton's second law *algebraically* to the rotational motion of the proton to find the speed of the proton.	
Assuming nonrelativistic speeds, use the speed to determine the maximum kinetic energy of the proton using the maximum possible radius.	$K_{\max} = 5.22$ MeV
Determine the potential difference across the dees by remembering that the protons are accelerated twice each revolution.	$\Delta V = 36\,000$ V

Check: The kinetic energy is much less than the 938 MeV rest energy of the proton, so the classical expression is appropriate. All the units work out properly.

Taking It Further: How would this problem change if the particles were electrons?

Try It Yourself #4

A beam of 2.50-MeV particles with charge $q = -2e$ is deflected by the magnetic field of a bending magnet as shown. The radius of curvature of the beam is 20.0 cm and the strength of the magnetic field is 1.50 T. What is the mass of the particles making up the beam?

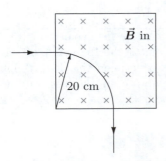

Picture: Apply Newton's second law to the particles, which will relate the radius, speed, charge, magnetic field, and mass. Find the speed from the kinetic energy of the particles.

Solve:

Use the kinetic energy to find an *algebraic* expression for the speed of the particles in terms of the energy and the mass.	
Algebraically apply Newton's second law to the circular motion of the particles.	
Substitute your result for the speed from the first step into the equation from the second step and solve for the mass.	$m = 1.15 \times 10^{-26}$ kg $= 6.94$ u

Check: This is a reasonable atomic mass, which corresponds to a Li^{2-} ion.

Taking It Further: Sketch the path for the given ions in the given magnetic field here as a reference path. Then sketch a path for a more massive doubly ionized negative ion and a less massive doubly ionized positive ion, both with the same speed as the Li ions, for comparison.

26.3   Torques on Current Loops and Magnets

In a Nutshell

A **current loop** is one or more complete turns of a current-carrying wire. The magnetic dipole moment $\vec{\mu}$ of a current loop is $\vec{\mu} = NIA\hat{n}$, where N is the number of turns, I the current in the wire, A the area of the plane surface bounded by a single turn of the wire, and $\hat{n}$ a unit vector normal to the surface. $\hat{n}$ points in the direction of the thumb when the fingers of the right hand curl in the direction of the current.

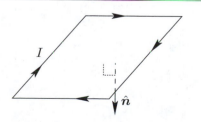

When a current loop is placed in a region containing a uniform magnetic field, there is no net force on the loop. However, the magnetic forces exerted on the loop produce a torque. The magnetic forces exerted by a uniform magnetic field on a current loop result in a couple, that is, a pair of oppositely directed offset forces of equal magnitude (for a discussion of couples see pages 405–406 of the text). The torque exerted by the field on a current loop is a vector given by $\vec{\tau} = \vec{\mu} \times \vec{B}$. This has exactly the same form as the torque exerted on an electric dipole $\vec{p}$ by an electric field $\vec{E}$.

A current loop in a magnetic field has some potential energy that depends on the orientation of the magnetic dipole moment $\vec{\mu}$ relative to $\vec{B}$. We define the zero point such that this energy is zero when the angle between $\vec{\mu}$ and $\vec{B}$ is 90°. The energy has a minimum when these vectors are oriented in the same direction, and a maximum when they are oriented in opposite directions. $U = -\vec{\mu} \cdot \vec{B}$.

Permanent magnets may also be described in terms of a magnetic dipole moment. Consequently, the discussions of torque and energy presented here can be applied not only to current loops but to permanent magnets as well.

Physical Quantities and Their Units

Magnetic dipole moment   SI units of $A \cdot m^2$

Important Derived Results

Magnetic dipole moment $\vec{\mu} = NIA\hat{n}$

Torque on a current loop $\vec{\tau} = \vec{\mu} \times \vec{B}$

Potential energy of a magnetic dipole $U = -\vec{\mu} \cdot \vec{B}$

Common Pitfalls

> ➤ In a uniform magnetic field a net torque, but *no* net force, is exerted on either permanent magnetic dipoles or current loops.

5. TRUE or FALSE: The torque exerted on a current loop by a uniform magnetic field tends to orient the plane of the loop at right angles to the magnetic field.

6. How is it possible for an object with no net force to experience a net torque?

Try It Yourself #5

There is a uniform magnetic field of 2.20 G in the positive x direction. A compass needle, a thin 2.00-cm-long iron rod with a cross-sectional area of 0.100 cm^2, is suspended from its center of mass so that it is free to rotate in the x,y plane. The needle is released from an initial position where it points in a direction making an angle of 135° with the positive x axis. As the needle rotates through the position where it is momentarily parallel with the y axis, its angular acceleration is 0.400 rad/s^2. What is its magnetic dipole moment?

Picture: The angular acceleration is due to the magnetic torque exerted on the magnet by the magnetic field. Use Newton's second law for rotations, in addition to the expression for magnetic torque, to determine the magnetic dipole moment.

Solve:

Draw a sketch of the compass needle, showing its initial and final positions relative to the uniform magnetic field.	
Write Newton's second law for rotations *algebraically* and substitute the *algebraic* expression for magnetic torque. Also substitute, *algebraically*, the moment of inertia for a rod rotated about its middle from Table 9-1 on page 295 of the text.	
Determine the mass of the rod. You may need to look up the density in Table 13-1 on page 424 of the text.	
Solve *algebraically* for the magnetic dipole moment. When the needle is parallel with the y axis, the magnetic moment is perpendicular to the magnetic field. Substitute values with their units to arrive at your final answer.	$\mu = 9.65 \times 10^{-5}$ A $\cdot$ m^2

Check: These values are reasonable and the units are correct.

Taking It Further: What is the potential energy of the compass needle when it is at its initial position?

Try It Yourself #6

A small 20-turn current loop with a 4.00-cm diameter is suspended in a region with a magnetic field of 1000 G, with the plane of the loop parallel with the magnetic field direction. What is the current in the loop when the torque exerted by the magnetic field on the loop is 4.00×10^{-5} N · m?

Picture: The torque on the current loop is related to its magnetic dipole moment. The magnetic dipole moment is due to current running through the current loop. When the plane of the current loop is parallel with the magnetic field, then the magnetic dipole moment of the current loop is perpendicular to the magnetic field.

Solve:

Draw a sketch of the loop in the magnetic field.	
Write an *algebraic* expression relating the torque to the magnetic dipole moment.	
Write an *algebraic* expression for the dipole moment of the coil in terms of the current, loop area, and number of turns.	
Substitute the result from the second step into the expression from the first step. Solve *algebraically* for the current in the loop.	

Substitute values with their units to get your final answer.	
	$I = 15.9$ mA

Check: This is a reasonable current and the units are correct.

Taking It Further: Describe the subsequent motion of the loop if it is allowed to rotate.

26.4 The Hall Effect

In a Nutshell

When a current-carrying wire is placed in a magnetic field $\vec{B}$, the magnetic force exerted on the charge carriers in the wire causes them to drift to one side of the wire. This results in a separation of charge across the wire, a phenomenon called the **Hall effect**. There is a potential difference V_H, called the **Hall voltage**, across the width of the wire due to the electric field $\vec{E}$ associated with this charge separation. $V_H = Ew = v_d Bw$, where v_d is the drift speed of the charge carriers, and w is the width of the wire.

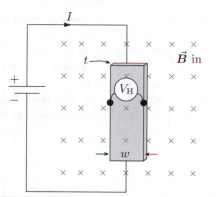

For a given magnetic field strength, the Hall voltage across a wire is proportional to the drift speed of the charge carriers. Thus, the Hall effect can be used to measure magnetic field strengths. For a known current, magnetic field strength, and wire width, the Hall effect can be used to determine the sign and number density of the charge carriers: $n = |I|B/(teV_H)$.

Important Derived Results

Hall voltage

$$V_H = Ew = v_d Bw$$

Number density of charge carriers

$$n = \frac{|I|B}{teV_H}$$

Common Pitfalls

> ➤ Although in metallic materials the charges actually moving are negatively charged electrons, this is not necessarily the case. In general either positive or negative charges can be responsible for the current.

7. TRUE or FALSE: Magnetic field lines always begin on the north pole of a magnet and end on the south pole.

8. The expressions for the Hall voltage can reduce to give the same expression that is used when analyzing the velocity selector. Explain why this makes sense.

Try It Yourself #7

As shown in the figure, the current is downward in a 4.00-cm-wide conductor with a rectangular cross section. A uniform magnetic field of 1.50 T is directed into the figure. The potential difference $V_b - V_a$ is +12.0 μV. (a) Are the charge carriers positively or negatively charged? (b) What is the drift speed of the charge carriers?

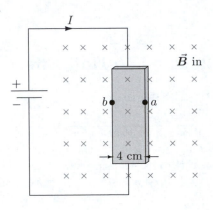

Picture: As charge carriers move through the rectangular wire, they experience a magnetic force which causes charge to migrate to the sides of the wire. These separated charges set up an electric field such that in the steady state, the electric field produced by these charges exerts a force on the subsequent charge carriers that directly cancels the magnetic force felt by the charges as they flow through the wire.

Solve:

$V_b > V_a$, which means that positive charges must have accumulated along side b as shown. Given the direction of the current, positive charge will flow "down" through the conductor, and negative charge will flow "up" through the conductor. Use the right-hand-rule to determine which charges must be moving for positive charges to accumulate on side b.	The charge carriers are negative.
Use the expression for the Hall voltage to determine the drift speed of the charge carriers.	$v_d = 2.00 \times 10^{-4}$ m/s

Check: This is a reasonable drift velocity.

Taking It Further: How does your answer change if everything remains the same in the problem except that the magnetic field is now out of the page?

Try It Yourself #8

A segment of 3.00-mm-diameter copper wire has a narrow region where the diameter is only 1.00 mm. This segment of current-carrying wire is in a region where there is a uniform magnetic field perpendicular to the wire. If the Hall voltage across the wire in the thin region is V_0, what is it across the thicker region?

Picture: Because the wire is continuous, the current and the number density of charge carriers throughout the wire is the same. Consequently, the current density, and hence the drift velocity, varies inversely with the cross-sectional area of the wire.

Solve:

Use the expression for the Hall voltage to solve *algebraically* for the drift speed of the charge carriers.	
Determine the drift speed in the thicker segment of wire.	
Use the value obtained in the second step to find the Hall voltage across the thicker wire segment.	$V_{\mathrm{H}} = V_0/9$

Check: This seems reasonable.

QUIZ

1. TRUE or FALSE: The Earth's south magnetic pole is near its north geographic pole.

2. TRUE or FALSE: For a given current, the magnetic force on a current-carrying wire is independent of the charge and drift speed of the charge carriers in the wire.

3. Both electric and magnetic fields can exert a force on a moving charge. In a particular case, how could you tell whether it is an electric or a magnetic force that is causing a moving charge to deviate from a straight-line path?

4. A velocity selector consists of crossed electric and magnetic fields, with the $\vec{B}$ field directed straight up. A beam of positively charged particles passing through the velocity selector from left to right is undeflected by the fields. (a) In what direction is the electric field? (b) The direction of the particle beam is reversed so that it travels from right to left. Is it deflected? If so, in what direction? (c) A beam of electrons (negatively charged) moving with the same speed is passed through from left to right. Is it deflected? If so, in what direction?

5. Physicists refer to crossed $\vec{E}$ and $\vec{B}$ fields as a velocity selector. In the same sense, the deflection of charged particles in a strong magnetic field perpendicular to their motion can be thought of as a momentum selector. Why is this?

6. A straight segment of wire 35.0 cm long carrying a current of 1.40 A is in a uniform magnetic field. The segment makes an angle of 53° with the direction of the magnetic field. If the force on the segment is 0.200 N, what is the magnitude of the magnetic field?

7. There is a uniform magnetic field of 1.20 T in the positive z direction. (a) A particle with a charge of -2.00 μC moves with a speed of 2.20 km/s in the y,z plane. Find the magnitude of the magnetic force on the particle when the direction of the velocity makes an angle of 55° with the positive y direction and 35° with the negative z direction. (b) A second particle, also with a charge of -2.00 μC, moves with a speed of 2.20 km/s in the x,y plane. Find the magnitude of the magnetic force on the particle when the direction of the velocity makes an angle of 55° with the positive y direction and an angle of 35° with the positive x direction.

Chapter 27

Sources of the Magnetic Field

27.1 The Magnetic Field of Moving Point Charges

In a Nutshell

Although many of the properties of permanent magnets have been known for 1000 years or more, it wasn't until the 1820s that Hans Christian Oersted observed that a current-carrying wire deflected a compass needle. This simple observation has led us to our contemporary understanding that currents are the source of magnetic field $\vec{B}$.

When a point charge q moves with velocity $\vec{v}$, it produces a magnetic field at some point P given by $\vec{B} = \dfrac{\mu_0}{4\pi} \dfrac{q\vec{v} \times \hat{r}}{r^2}$. In this expression $\hat{r}$ is the unit vector of $\vec{r}$, which points *from* the moving charge *toward* the field point P. The quantity $\mu_0 = 4\pi \times 10^{-7}$ T·m/A is known as the **magnetic constant** or **permeability of free space**.

Physical Quantities and Their Units

Permeability of free space $\qquad\qquad\qquad\qquad \mu_0 = 4\pi \times 10^{-7}$ T · m/A $= 4\pi \times 10^{-7}$ N/A^2

Fundamental Equations

Magnetic field of a point charge $\qquad\qquad\qquad \vec{B} = \dfrac{\mu_0}{4\pi} \dfrac{q\vec{v} \times \hat{r}}{r^2}$

Common Pitfalls

> Most students have difficulty visualizing the three-dimensional aspect of magnetism. You can help your visualization by making drawings of every situation. Most istructors practice these drawings before coming to class, and you may want to do the same. If appropriate, make several drawings, each from a different point of view. It is often helpful to use different colors to illustrate cross products.

> You can find the directions of the magnetic forces that two moving charges exert on each other as follows. First, at the location of one charge, find the direction of the magnetic field due to the other charge; and repeat the process at the location of the other charge. Second, find the directions of the forces exerted by these magnetic fields. Use the same two steps to find the directions of the forces exerted by two current elements on each other (you can practice this in the next section).

1. TRUE or FALSE: If a bullet moving directly away from an observer acquires a negative charge, the magnetic field lines due to this moving charge are directed counterclockwise as viewed by the observer.

2. Discuss the similarities between the expressions for $\vec{E}$, the electric field created by a point charge and $\vec{B}$, the magnetic field created by a moving point charge.

Try It Yourself #1

A particle with a charge of +5.00 nC travels through the origin with a velocity of $(3.00 \times 10^6 \text{ m/s})\hat{k}$. Find the magnitude and direction of the magnetic field due to this charged particle at points $a = (0, 0, 5.00 \text{ cm})$, $b = (0, 0, -5.00 \text{ cm})$, $c = (0, 5.00 \text{ cm}, 0)$, $d = (5.00 \text{ cm}, 0, 0)$, $e = (3.00 \text{ cm}, 4.00 \text{ cm}, 0)$, and $f = (3.00 \text{ cm}, 0, 4.00 \text{ cm})$.

Picture: Sketch the situation and use the formula for the magnetic field due to a charged particle to find the field in each case.

Solve:

Make a sketch of the moving particle and the locations where the magnetic field is to be located. The right-hand rule can be used to determine the direction of the magnetic field and to check your calculations.	
Write the general *algebraic* expression for the magnetic field due to a moving point charge to use as a reference.	
For all the points in this problem, the distance r from the moving charge to the point of interest is 5 cm, so we can simplify the calculations by finding the numerical value of $\mu_0 qv/(4\pi r^2)$ just once. Then for each point all we have to do is evaluate $\hat{v} \times \hat{r}$, where $\hat{v} = \hat{k}$ is the unit vector of the velocity.	$\dfrac{\mu_0 qv}{4\pi r^2} = 6.00 \times 10^{-7}$ T
Determine the value of $\hat{r}$ for point a, evaluate the cross product mathematically, and determine the magnetic field at this point.	$\vec{B}_a = 0$ T

Determine the value of $\hat{r}$ for point b, evaluate the cross product mathematically, and determine the magnetic field at this point.	$\vec{B}_b = 0$ T
Determine the value of $\hat{r}$ for point c, evaluate the cross product mathematically, and determine the magnetic field at this point.	$\vec{B}_c = (6.00 \times 10^{-7} \text{ T})(-\hat{\imath})$
Determine the value of $\hat{r}$ for point d, evaluate the cross product mathematically, and determine the magnetic field at this point.	$\vec{B}_d = (6.00 \times 10^{-7} \text{ T})\hat{\jmath}$
Determine the value of $\hat{r}$ for point e, evaluate the cross product mathematically, and determine the magnetic field at this point.	$\vec{B}_e = (6.00 \times 10^{-7} \text{ T})(-0.8\hat{\imath} + 0.6\hat{\jmath})$
Determine the value of $\hat{r}$ for point f, evaluate the cross product mathematically, and determine the magnetic field at this point.	$\vec{B}_f = (3.60 \times 10^{-7} \text{ T})\hat{\jmath}$

Check: The right-hand rule for cross products can be used to check our directions. The units all are correct.

Taking It Further: Why is the magnetic field zero for any point directly along the instantaneous trajectory of the particle?

Try It Yourself #2

Particle 1 with a charge of $+5.00$ nC travels through the origin with a velocity of $(1.40 \times 10^6 \text{ m/s}) \, \hat{\boldsymbol{k}}$. At the same instant, particle 2, with a charge of $+2.00$ nC, travels through the point $(5.00 \text{ cm}, 0, 0)$ with a velocity of $(-3.00 \times 10^6 \text{ m/s})\hat{\boldsymbol{i}}$. At that instant, find (a) the magnetic force exerted by the field of particle 1 on particle 2 and (b) the magnetic force exerted by the field of particle 2 on particle 1.

Picture: Note that these are *not* Newton's third-law force pairs. The particles are not exerting forces directly on each other! The particles create fields, and it is these fields that exert a force on each particle. First find the field created by one of the particles, and then determine the magnetic force experienced by the second moving particle while under the influence of the field of the first particle.

Solve:

Draw separate sketches for parts (a) and (b).	
Determine the magnetic field created by particle 1 at the position of particle 2.	
Determine the force on particle 2 when it experiences the magnetic field calculated in the previous step.	$\vec{\boldsymbol{F}} = (-1.68 \times 10^{-9} \text{ N})\hat{\boldsymbol{k}}$
Determine the magnetic field created by particle 2 at the position of particle 1.	

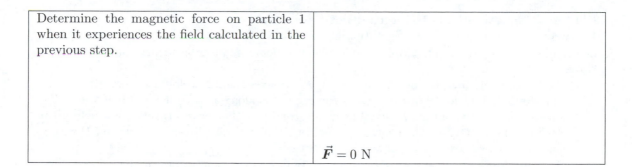

Determine the magnetic force on particle 1 when it experiences the field calculated in the previous step.	
	$\vec{F} = 0$ N

Check: Use the right-hand rule and the definition of cross product to double-check these results. In order to make sense of this apparent paradox (there appears to be a net force on this system of two charged particles, which means a net change in momentum of the system in the absence of an external force) we need to consider the momentum that is carried by the electric and magnetic fields involved in this problem. We will see how to think properly about this concept in Chapter 30.

Taking It Further: In the meantime, use the space provided here to convince yourself that two charges moving parallel or antiparallel to each other *do* experience equal and opposite forces, despite the fact they are not third-law force pairs.

27.2   The Magnetic Field of Currents: The Biot-Savart Law

In a Nutshell

If the source term $q\vec{v}$ in the expression for the magnetic field of a point charge is replaced by the equivalent current element $I\,d\vec{\ell}$, as was done in Chapter 26, we arrive at an expression for the magnetic field of a current element: $d\vec{B} = \dfrac{\mu_0}{4\pi}\dfrac{I\,d\vec{\ell} \times \hat{r}}{r^2}$. This relationship is known as the **Biot-Savart law**, and it is analogous to Coulomb's law for the electric field due to a small charge element dq. This expression can be used to calculate the magnetic field due to a variety of circuits.

Finding Magnetic Fields Using the Biot-Savart Law

Picture: Sketch the current configuration along with a field point P (the point where $\vec{B}$ is to be calculated). The sketch should include a current element $d\vec{\ell}$ at an arbitrary source point S.

Solve:

1. Add coordinate axes to the sketch. The choice of axes should exploit any symmetry of the current configuration. For example, if the current is along a straight line, then select that line as one of the coordinate axes. Draw a second axis that passes through the field point P. In addition, include the coordinates of both P and S, the distance r between P and S, and the unit vector $\hat{r}$ directed away from S toward P.

2. To compute the magnetic field $\vec{B}$ by integration, we express $d\vec{B}$ in component form. First perform the cross product $d\vec{\ell} \times \hat{r}$ to arrive at the individual components for $d\vec{B}$.

3. Integrate each nonzero component of the magnetic field separately to find the net magnetic field in each component direction.

Check: Always use the right-hand rule to double-check your calculation of the resulting direction of $\vec{B}$. If they don't agree, double-check your work.

Direct integration can be used to find the magnitude of the magnetic field some distance z along the axis of a current-carrying loop. $B_z = \dfrac{\mu_0}{4\pi} \dfrac{2\pi R^2 I}{(z^2 + R^2)^{3/2}}$, where I is the current in the loop and R is the radius of the loop. The right-hand rule is used to determine the direction of $\vec{B}$.

A **solenoid** is conducting wire wound into a helix of closely spaced turns, as shown. The magnetic field along the axis of the helix is nothing more than the sum (integral) of the field from each individual current loop. If the solenoid is relatively short, then $B_z =$

$$\frac{1}{2}\mu_0 nI \left(\frac{z - z_1}{\sqrt{(z - z_1)^2 + R^2}} - \frac{z - z_2}{\sqrt{(z - z_2)^2 + R^2}} \right),$$

where I is the current in the solenoid, R is the radius of the solenoid, and n is the number density of turns (the number of turns per unit length along the axis of the solenoid). If the solenoid is "long" (its length is much larger than its radius) then this expression reduces to $B_z = \mu_0 nI$. Again, the right-hand rule is used to determine the direction of $\vec{B}$.

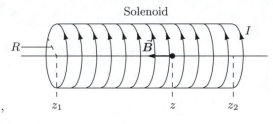

Solenoid

Some distance R from an infinitely long wire, the magnitude of the magnetic field is $B = \dfrac{\mu_0 I}{2\pi R}$. The right-hand rule is used to determine the direction of the magnetic field.

The **ampere** is defined as "that constant current which, if maintained in two straight, parallel conductors of infinite length and of negligible circular cross sections placed one meter apart in a vacuum, would produce a force between the conductors equal to 2×10^{-7} newtons per meter of length."

Fundamental Equations

Biot-Savart law
$$d\vec{B} = \frac{\mu_0}{4\pi} \frac{I\,d\vec{\ell} \times \hat{r}}{r^2}$$

Important Derived Results

Magnetic field of a current loop	$B_z = \dfrac{\mu_0}{4\pi} \dfrac{2\pi R^2 I}{(z^2 + R^2)^{3/2}}$
Magnetic field inside a solenoid	$B_z = \dfrac{1}{2}\mu_0 n I \left(\dfrac{z - z_1}{\sqrt{(z - z_1)^2 + R^2}} - \dfrac{z - z_2}{\sqrt{(z - z_2)^2 + R^2}} \right)$
Magnetic field inside a "long" solenoid	$B_z = \mu_0 n I$
Magnetic field of an infinitely long wire	$B = \dfrac{\mu_0 I}{2\pi R}$

Common Pitfalls

> ➤ In can be relatively easy to become confused by the many right-hand rules useful for magnetics. If you remember only one, make sure it is the right-hand rule for the direction of a cross product of two vectors. Although not always the simplest, this right-hand rule works for any situation you can imagine as long as your properly identify the current element $I\,d\vec{\ell}$ and $\hat{r}$.

> ➤ The right-hand rules for current loops are the following. If you wrap the fingers of your right hand in the direction of the current loop, your thumb points in the direction of (a) the on-axis magnetic field of the loop; (b) the magnetic dipole moment $\vec{\mu}$ of the loop; (c) the positive direction of the normal vector to the loop $\hat{n}$, and (d) the direction of the north magnetic pole of the current loop.

> ➤ Long wires also have a right-hand rule. If you point the thumb of your right hand in the direction of the current element $I\,d\vec{\ell}$, the fingers of your right hand will wrap around the wire in the direction of the magnetic field created by the wire.

3. TRUE or <u>FALSE</u>: On the surface of a sphere centered on a current element, the magnitude of the magnetic field due to the current element is constant. *B varies as sinθ which range from θ = 0 to θ = π*

4. In the figure a mass hangs in equilibrium on the end of a spring. If a current is passed through the spring, which way does the mass move? Explain. *Attractive force cause contraction — parallel wire*

Try It Yourself #3

Two long, straight, current-carrying wires in the x,y plane are parallel to the y axis and intersect the x axis at $x = \pm 4.00$ cm, respectively. Each wire carries a constant current of 20.0 A in the positive y direction. Find the magnetic field at point P, which is at $z = 3.00$ cm on the z axis.

Picture: Use the formula for the field due to a long straight wire to find the contribution from each wire to the magnetic field at point P. Add the magnetic field vector of each contribution to get the net magnetic field at P.

Solve:

Sketch the physical situation. Let the page be the x,z plane, with the $+x$ axis to the right, the $+z$ axis up, and the $+y$ axis into the page. Label the current-carrying wires 1 and 2, respectively.	
Determine the magnitude of the magnetic field produced by each current at point P using the formula for the field due to a long straight wire.	
The directions of the magnetic field from each wire can be determined by the right-hand rule. Be precise when determining this direction. It may help to draw on your figure the distance r from each wire and remember that the magnetic field must be perpendicular to this line and the current. Draw a vector for the magnetic field due to each wire. Make sure your label them properly. It can be easy to confuse which $\vec{B}$ is due to which wire. Using the geometry of the problem, determine the vector components of $\vec{B}$ for the field created by each wire.	
Sum the two magnetic fields to find the net magnetic field at P.	$\vec{B} = \left(9.60 \times 10^{-5} \text{ T}\right)\hat{\imath}$

Check: Because of the symmetry of the problem, the $\hat{k}$ components of the magnetic fields will cancel at point P, and the $\hat{\jmath}$ components must be zero because the magnetic field must be perpendicular to the current.

Taking It Further: What will be the magnetic field at $z = -3.00$ cm on the z axis? Explain, using symmetry arguments and the results of your calculations above.

Try It Yourself #4

Find the magnetic field at point P in the figure. The curved section is $\frac{3}{4}$ of a 15.0-cm-radius circle centered at point P. The current is a constant 40.0 A.

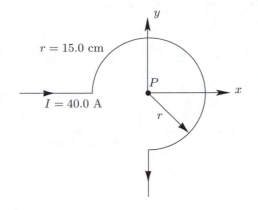

Picture: The magnetic field is the vector sum of the fields due to each straight segment plus the field due to the curved segment. Apply the Biot-Savart law to each segment.

Solve:

Draw a sketch splitting the wire into three segments, two straight and one curved.	
Determine the magnetic field at P due to the incoming straight current segment by applying the Biot-Savart law. Make sure to take advantage of the angle between $I\,d\vec{\ell}$ and $\hat{r}$.	
Determine the magnetic field at P due to the outgoing straight current segment by applying the Biot-Savart law. Make sure to take advantage of the angle between $I\,d\vec{\ell}$ and $\hat{r}$.	
Determine the magnetic field at P due to the curved segment. Begin by identifying a current element on the curved segment and writing $d\vec{\ell}$ in component form as a function of θ, the angle measured from the $-x$ direction. Be careful of your signs.	
Now draw an $\vec{r}$ from your current element to P and write $\vec{r}$ in component form, also as a function of θ.	

From $\vec{r}$ find both $\hat{r}$ and r for your current element.	
Now perform the cross product $d\vec{\ell} \times \hat{r}$.	
Substitute this result into the Biot-Savart law, finding an expression for $d\vec{B}$, the differential magnetic field due to your current element. Finally, integrate each component of $d\vec{B}$ to find the magnetic field at point P due to the curved segment. Watch your limits of integration, as the angle θ is defined in a nonstandard way.	
Sum the magnetic field due to each of the three segments to find the total magnetic field at P.	$\vec{B} = (1.26 \times 10^{-4} \text{ T})(-\hat{k})$

Check: Use the right-hand rule for current loops to double-check the direction of the magnetic field.

Taking It Further: It turns out that this result is simply $\frac{3}{4}$ of the magnetic field from a full current loop. However, the calculation was required to confirm this. Given the symmetry of this problem, describe here how and why you could have done this problem without determining $d\vec{\ell}$ and $\vec{r}$ explicitly and mathematically carrying out the cross product. These kinds of symmetry arguments are extremely powerful but are not always applicable, which is why it is important for you to be able to carry out all the steps outlined above.

27.3   Gauss's Law for Magnetism

In a Nutshell

Gauss's law for magnetism is a mathematical statement of the fact that humans have not yet discovered an isolated magnetic monopole. Because magnetic sources always come as dipoles, every magnetic field line that penetrates a closed surface in the "outward" direction must *always* wrap around and penetrate the close surface in the "inward" direction as well.

Fundamental Equations

Gauss's law for magnetism
$$\oint_S \vec{B} \cdot \hat{n}\, dA = \oint_S B_n\, dA = 0$$

27.4   Ampère's Law

In a Nutshell

The integral form of **Ampère's law** is $\oint_C \vec{B} \cdot d\vec{\ell} = \oint_C B_t\, d\ell = \mu_0 I_C$, where C is any closed curve and I_C is the current that passes through the surface S bounded by C.

Ampère's law is valid for any situation in which the currents are steady and continuous. However, it is *useful* only in highly symmetric situations that allow us to easily evaluate the integral side of the expression. In particular, Ampère's law only gives the component of $\vec{B}$ parallel to $d\vec{\ell}$ around the curve B_t. In addition, sufficient symmetry should be present to ensure that B_t is uniform over the entire $d\ell$. *Only in these circumstances*, Ampère's law reduces to $B_t \ell = \mu_0 I_C$, from which B_t can be readily calculated from I_C.

Ampère's law and the Biot-Savart law can be used only to determine magnetic fields produced by circuits in which both the current and the charge distribution are constant in time. These laws do provide close approximations to actual magnetic fields produced by circuits in which time-varying changes take place sufficiently slowly. Magnetic fields produced by circuits in which the current or the charge distribution changes in time will be studied in Chapter 28.

Fundamental Equations

Ampère's law
$$\oint_C \vec{B} \cdot d\vec{\ell} = \oint_C B_t\, d\ell = \mu_0 I_C$$

Important Derived Results

B inside a tightly wound toroid
$$B = \frac{\mu_0 N I}{2\pi r}, \text{ for } a < r < b$$

Common Pitfalls

> Ampère's law and the Biot-Savart law can be used only for circuits in which the current and the charge distribution are constant in time. These laws provide close approximations to the magnetic field produced by circuits in which changes in the current and the charge distribution take place sufficiently slowly.

5. TRUE or FALSE: In principle, Ampère's law can be used only in situations of sufficient symmetry that the direction of the magnetic field can be deduced from symmetry arguments.

6. Think of a long solenoid of N total turns carrying current I as a helical spring. If the solenoid is stretched along its length without changing N or I, does the field inside it change? How about its magnetic moment? Explain.

Try It Yourself #5

A long, straight, thin-walled cylindrical shell of radius R carries a steady current I flowing parallel with the shell's axis. Find B at all points inside the shell and outside the shell.

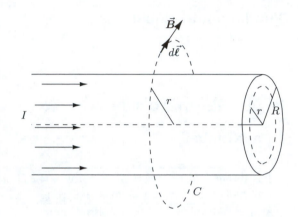

Picture: There is a significant amount of symmetry in this problem, so Ampère's law should be used. There are two regions in this problem, radii less than R and radii greater than R. Ampère's law will be applied differently for these two regions.

Solve:

Apply Ampère's law to the closed loop of radius $r < R$, often called an **ampèrian loop** because it is used in Ampère's law, to find the magnetic field in that region. Because of symmetry, $\vec{B}$ is parallel to $d\vec{\ell}$ everywhere along the loop of radius r, which eliminates the dot product. In addition, because of symmetry the magnitude of $\vec{B}$ is constant everywhere along the loop of radius r because of symmetry, so that B can be pulled outside the integral, leaving the left-hand side of the equation equal to the magnitude of the magnetic field times the circumference of the ampèrian loop. Determine the amount of current that goes through the loop of radius $r < R$ and solve for B.	$B = 0$ for $r < R$
For the situation with the ampèrian loop outside of the conducting shell, $r > R$, the same symmetry arguments can be made. Determine the amount of current that goes through the loop of radius $r > R$ and solve for B.	$B = \dfrac{\mu_0 I}{2\pi r}$ for $r > R$

Check: Regardless of the symmetric charge distribution, outside a wire the magnetic field looks like that of an infinitely long wire with total current I.

Taking It Further: Could you use Ampère's law to calculate the magnetic field if the wire were relatively short? Explain.

Try It Yourself #6

A long, straight, coaxial cable consists of a solid wire of radius a and a thin-walled conducting coaxial shell of radius $b > a$. The wire carries a steady current I uniformly distributed in one direction, and the shell carries the same current in the opposite direction. Find B at all points inside the wire, between the wire and the shell, and outside the shell.

Picture: There is a significant amount of symmetry in this problem, so Ampère's law should be used. There are three regions in this problem, so you will need to apply Ampère's law differently for these three regions, depending on how much current is enclosed.

Solve:

Draw a sketch of the situation, with three ampèrian loops, one for each region of interest.	
Determine the amount of current that flows through a loop of radius $r < a$. It will not simply be I.	
Now that you know how much current flows through the loop, apply Ampère's law to the loop of radius $r < a$ to find the magnetic field in that region. Remember to make the appropriate symmetry arguments.	$B = \dfrac{\mu_0 I r}{2\pi a^2}$ for $0 \leq r < a$

Apply Ampère's law to the region $a < r < b$.	
	$B = \dfrac{\mu_0 I}{2\pi r}$ for $a < r < b$
Apply Ampère's law to the region $r > b$.	
	$B = 0$ for $r > b$

Check: Since the currents are equal and opposite, for an ampèrian loop completely outside the cable the current sums to zero, which for symmetric situations means that $B_t = 0$ also.

Taking It Further: Would it possible to change the conditions of this problem so that the magnetic field for $r > b$ would be in the opposite direction of the magnetic field in the region $a < r < b$? Explain.

27.5   Magnetism in Matter

In a Nutshell

Most materials fall into one of three main magnetic categories—ferromagnetic, paramagnetic, or diamagnetic—according to the manner in which their atoms or molecules align themselves in an external magnetic field. Most atoms and molecules have an intrinsic dipole moment that is due to both the orbital motion and the spin of their electrons; the atoms in these molecules tend to align their magnetic dipole moments with magnetic fields. Materials consisting of these atoms are said to be either **paramagnetic** or **ferromagnetic**. Some molecules lack intrinsic magnetic moments; for such molecules a magnetic field induces a dipole moment that is directed oppositely to the field direction. Materials consisting of such molecules are said to be **diamagnetic**.

The degree to which a material is magnetized depends on the individual strengths of the molecular magnetic dipole moments and the degree to which they are aligned. The **magnetization $\vec{M}$** of a material is the net magnetic dipole moment per unit volume of the material: $\vec{M} = d\vec{\mu}/dV$.

Magnetic dipoles can be thought of as current loops. In a magnetized material in which the magnetic dipole moments are aligned, these current loops are distributed throughout the material. If the material is uniformly magnetized, the currents in neighboring current loops cancel. However, at the edge of the material there are no neighboring current loops on one side, and so there is a net current, called an **ampèrian current**, on the outside edge of the material as in the figure. The ampèrian current is due to the motions of bound rather than free charges. It can be shown (see page 938 of the text) that the magnetic field inside a long bar magnet is given by $\vec{B} = \mu_0 \vec{M}$.

Surface current

Consider an unmagnetized cylinder of material that is located within a solenoid. When a current flows through the windings of the solenoid, a magnetic field $\vec{B}$ fills the region within it. This applied field of the solenoid magnetizes the material of the cylinder so that it has magnetization $\vec{M}$. There are now two magnetic fields in the material, an applied field $\vec{B}_{\text{app}}$ due to the current in the solenoid and a second field due to the magnetization of the material. The net magnetic field $\vec{B}$ is the sum of the two—that is, $\vec{B} = \vec{B}_{\text{app}} + \mu_0 \vec{M}$.

If the atoms constituting the cylinder have intrinsic magnetic dipole moments, $\vec{B}_{\text{app}}$ exerts torques that tend to align them with it. For these ferromagnetic and paramagnetic materials, $\vec{M}$ is parallel with $\vec{B}_{\text{app}}$. If the atoms of the cylinder do not have intrinsic magnetic dipole moments, $\vec{B}_{\text{app}}$ induces a dipole moment that is opposite in direction to it. For these diamagnetic materials, $\vec{M}$ is opposite to $\vec{B}_{\text{app}}$. For both paramagnetic and diamagnetic materials, the magnetization is proportional to the applied field, Thus, $\vec{M} = \chi_{\text{m}} \left(\vec{B}_{\text{app}}/\mu_0 \right)$, where the dimensionless constant χ_{m}, called the **magnetic susceptibility**, is a property of the material. Substituting this expression for $\vec{M}$ in the previous equation, we have $\vec{B} = \vec{B}_{\text{app}}(1 + \chi_{\text{m}})$. The quantity $K_{\text{m}} = 1 + \chi_{\text{m}}$ is called the **relative permeability** of the material.

For paramagnetic materials, χ_{m} is a small positive number that depends on temperature. For diamagnetic materials, it is a small negative number that is independent of temperature. The magnetic susceptibility of various materials are listed in Table 27-1 on page 939 of the text.

The fundamental constant h (with units of angular momentum), called **Planck's constant**, has a value of 6.67×10^{-34} J·s, and the combination $\hbar = h/(2\pi) = 1.05 \times 10^{-34}$ J·s, is read "h bar."

The magnetic dipole moment $\vec{\mu}$ of a particle with mass m and charge q moving in a circular orbit can be shown to be $\vec{\mu} = q\vec{L}/(2m)$, where $\vec{L}$ is the angular momentum of the particle about the center of the circle. For an electron, $m = m_e$ and $q = -e$, so the magnetic moment of the electron due to its orbital motion is $\vec{\mu}_\ell = -\mu_{\text{B}}\vec{L}/\hbar$, where $\mu_{\text{B}} = e\hbar/(2m_e) = 9.27 \times 10^{-24}$ A·m^2 is the quantum unit of magnetic moment, called a **Bohr magneton**.

The magnetic moment of an electron due to its intrinsic spin angular momentum $\vec{S}$ is $\vec{\mu}_S = -2\mu_B\vec{S}/\hbar$.

Although the calculation of the magnetic moment of any atom is a complicated problem in quantum theory, the result for all atoms, according to both theory and experiment, is that the magnetic moment is of the order of a few Bohr magnetons. For atoms with zero net angular momentum, the net magnetic moment is zero. Angular momentum is found only in certain discrete values; that is, angular momentum is quantized.

Magnetization is greatest when all the atomic magnetic dipole moments are maximally aligned. For this case, the magnetization is called the **saturation magnetization** $M_{\text{s}} = n\mu$, where n is the number of molecules per unit volume and μ the magnetic dipole moment of each molecule.

Paramagnetic materials have a very small, positive magnetic susceptibility. The atoms or molecules in a paramagnetic material have an intrinsic magnetic dipole moment. An external magnetic field applied to a paramagnetic material exerts torques on the molecules that tend to align the dipole moments with the field. Not all the molecules are perfectly aligned with the field because thermal agitation keeps jarring them out of alignment. Typical values for the magnetic susceptibility of paramagnetic solids are on the order of 10^{-5}. The magnetization in weak magnetic fields varies inversely with temperature according to **Curie's law**: $M = \mu B_{\text{app}} M_{\text{s}}/(3kT)$.

Ferromagnetic materials have very large, positive values of magnetic susceptibility χ_{m}. In these materials, a small applied external magnetic field can produce a very high degree of alignment of the atomic magnetic dipole moments. This is because, unlike the atoms of paramagnetic materials, the magnetic dipole moments of the atoms of ferromagnetic materials interact strongly and form regions, called **magnetic domains**, in which they align with one another even when no applied field is present. At temperatures above a critical temperature called the **Curie temperature**, thermal agitation is sufficient to break up the domains and to prevent new ones from forming.

When an unmagnetized ferromagnetic material is located within a solenoid, the magnetization of the material increases as the solenoid current, and thus the applied magnetic field, increases. The magnetization, however, does not increase linearly with the applied field because the magnetization eventually approaches saturation for the material. If, after saturation is reached, the applied magnetic field is returned to zero, the material remains magnetized with some residual magnetization. This effect is called **hysteresis**. The magnetic field due to the residual magnetism is called the **remnant field**.

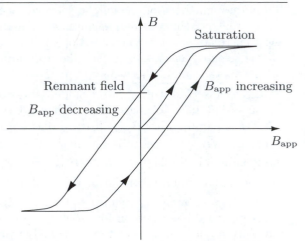

Ferromagnetic materials in which hysteresis effects are small are said to be **magnetically soft**. Magnetically soft materials are desirable as transformer cores. For permanent magnets, large hysteresis effects are desirable; materials in which hysteresis effects dominate behavior are said to be **magnetically hard**.

Diamagnetic materials have very small, negative values of magnetic susceptibility χ_{m}. An applied magnetic field induces currents in the atoms and molecules of any material. In accordance with Lenz's law, in a diamagnetic material these currents produce a magnetic flux that opposes the increase of the applied field, as we shall see in Chapter 28. Therefore, the induced magnetic dipole moments are anti-aligned with the applied field. The atoms or molecules of diamagnetic materials have no intrinsic magnetic dipole moment, so the magnetization of these materials is due entirely to the induced magnetic dipole moments. Thus the magnetic susceptibility of diamagnetic materials is negative.

A superconductor is a perfect diamagnetic material. An applied magnetic field induces currents in a superconductor that decrease the net magnetic field inside the superconductor to zero. Thus the magnetic susceptibility of a superconductor is $\chi_{\text{m}} = -1$.

Physical Quantities and Their Units

Bohr magneton	$\mu_{\text{B}} = \dfrac{e\hbar}{2m_e} = 9.27 \times 10^{-24} \text{ A} \cdot \text{m}^2 = 9.27 \times 10^{-24} \text{ J/T}$	
Planck's constant	$h = 2\pi\hbar = 6.67 \times 10^{-34} \text{ J} \cdot \text{s}$	

Important Derived Results

Definition of magnetization $\qquad\qquad\qquad\qquad \vec{M} = \dfrac{d\vec{\mu}}{dV}$

Orbital magnetic moment $\qquad\qquad\qquad\quad \vec{\mu} = \dfrac{q}{2m}\vec{L}$

Curie's law $\qquad\qquad\qquad\qquad\qquad\qquad M = \dfrac{1}{3}\dfrac{\mu B_{\mathrm{app}}}{kT}M_{\mathrm{s}}$

Definition of permeability $\qquad\qquad\qquad \mu = K_{\mathrm{m}}\mu_0 = (1 + \chi_{\mathrm{m}})\mu_0$

Common Pitfalls

> The term *magnetization* doesn't sound like a density, but it is. It is the magnetic dipole moment per unit volume—that is, the magnetic dipole moment density.
> The induced magnetic field far exceeds the applied field in ferromagnetic materials.
> The transition from a ferromagnetic material to a paramagnetic material when the temperature crosses the Curie point is not a gradual one. Rather, it is similar to the transition of water from liquid to gas when the temperature crosses the steam point. Such abrupt transitions are called *phase transitions*.
> Atoms and molecules of diamagnetic materials have induced magnetic dipole moments that are always opposed to the field direction. Atoms and molecules of paramagnetic and ferromagnetic materials have permanent magnetic dipole moments that are not 100 percent aligned except at saturation.
> All materials have a diamagnetic effect. The effect is so small, however, that is generally swamped by the much larger paramagnetic or ferromagnetic effects when they are present.

7. TRUE or FALSE: A diamagnetic material is attracted by either pole of a magnet.

8. Explain why a magnet will pick up an unmagnetized iron nail but not an otherwise identical aluminum nail.

Try It Yourself #7

The 1500-turn winding of a 1.00-cm-diameter solenoid 20.0 cm long carries a current of 1.00 A. The inside of the solenoid is filled with tungsten at 20.0°C. What are B_{app}, M, and B at the center of the solenoid?

Picture: The applied field is provided by the solenoid. The susceptibility of tungsten can be found in Table 27-1 on page 939 of the text. The magnetization is related to the susceptibility and adds a component to the total magnetic field inside the solenoid.

Solve:

Determine the magnetic field produced by the solenoid.	$B_{\text{app}} = 9.43 \times 10^{-3}$ T
Determine the magnetization of the tungsten from its susceptibility and the applied field.	$M = 0.510$ A/m
Determine the total magnetic field inside the solenoid.	$B = 9.43 \times 10^{-3}$ T

Check: Tungsten is a paramagnetic material, so we expect only a small effect on the magnetic field, which is what was calculated.

Taking It Further: How would this problem change if the tungsten were replaced by an unmagnetized iron bar?

Try It Yourself #8

Aluminum has a density of 2700 kg/m^3 and a molar mass of 0.0270 kg/mol. What is the magnetic dipole moment of an aluminum atom in Bohr magnetons?

Picture: The magnetic susceptibility of aluminum can be found in Table 27-1 on page 939 of the text. Note that it is small and positive, which tells us that aluminum is paramagnetic. This means we can use Curie's law, which relates the atomic magnetic dipole moment to other variables.

Solve:

Write Curie's law.	
Write the expression used to define magnetic susceptibility.	
Write the expression for the saturation magnetization in terms of the mass density, the molar mass, Avogadro's number, and the dipole moment.	
Divide the first expression by the second and substitute for M_s using the third expression.	
Solve the above expression for the magnetic moment μ.	
Convert this magnetic moment into Bohr magnetons.	$\mu = 0.209\mu_B$

Check: Because aluminum is paramagnetic, we expect it to have a small (< 1) magnetic moment, which we found.

QUIZ

1. TRUE or FALSE: The magnetic field lines associated with the magnetic field due to the steady current in a long straight wire are circles centered on the wire.

2. TRUE or FALSE: When the magnetic dipole moment of a current loop is aligned in an external magnetic field, the magnetic field that is due to the current loop and that is aligned along the loop's axis is also aligned with the external field.

3. Two flat, square, current-carrying loops lying in the plane of the page are shown in the figure. Explain how the following statements about magnetic forces apply to this case: Like poles repel, whereas opposite poles attract; parallel currents attract, whereas antiparallel currents repel.

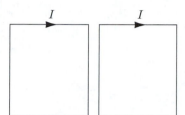

4. A permanent magnet may lose much of its magnetization if it is dropped or banged against something. Why?

5. What are the similarities and differences between the fields of electric and magnetic dipoles? Sketch the fields of an electric dipole and a current loop to illustrate your responses.

6. A long iron-core solenoid of carries a current of 10.0 mA. At this current the relative permeability is 1200. (a) What is the magnetic field strength within the solenoid? (b) Find the current necessary to produce the same field in the solenoid with the iron core removed.

7. In the figure, find the magnetic field at point *P*. The current is 80.0 A.

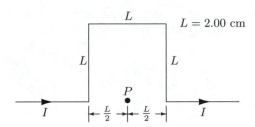

Chapter 28

Magnetic Induction

28.1 Magnetic Flux

In a Nutshell

The **magnetic flux** ϕ_m through a surface S is equal to the integral of the normal component B_n of the magnetic field on the surface times the area dA of each element of surface area. That is, $\phi_m = \int_S B_n \, dA = \int_S \vec{B} \cdot \hat{n} \, dA$, where $\hat{n}$ is the unit normal to the surface. The SI unit of flux is the **weber** (Wb).

Physical Quantities and Their Units

Magnetic flux   SI unit of weber, $1\ \text{Wb} = 1\ \text{T} \cdot \text{m}^2$

Important Derived Results

Magnetic flux $$\phi_m = \int_S B_n \, dA = \int_S \vec{B} \cdot \hat{n} \, dA$$

Common Pitfalls

> ➤ The sign of the magnetic flux is ambiguous. There are always two normals to a surface. You need to explicitly define the positive normal for each problem.

Try It Yourself #1

A ring of radius 8.00 cm lying in the x,y plane is centered at the origin. A magnetic field given by $\vec{B} = (6.00\ \text{T}) \cos\theta \, \hat{k}$, with θ measured with respect to the $+x$ axis but within the x,y plane, is also present. What is the magnitude of the flux through the ring?

Picture: Use the definition of magnetic flux to find the flux in this case.

Solve:

Draw a sketch of the situation, making sure to label the coordinate axes and the angle θ.	

Write the definition of magnetic flux as a guide.	
Determine the direction of $\hat{n}$, the normal to the surface.	
Find an *algebraic* expression for the differential area of the ring, in terms of θ.	
Substitute expressions and calculate the dot product $\vec{B} \cdot \hat{n}\,dA$.	
Now identify your limits of integration on θ and integrate to find the total magnetic flux.	
	$\phi_{\mathrm{m}} = 0$

Check: Using symmetry arguments, convince yourself that this is the correct answer.

Taking It Further: What is the magnitude of the magnetic flux through the half of the circle where $x > 0$? What about the half where $y > 0$? Explain.

28.2 Induced EMF and Faraday's Law

In a Nutshell

In the early 1830s, Michael Faraday and Joseph Henry independently discovered, by direct observation, that a *changing* magnetic flux through a surface induced an emf around the boundary of that surface. The emf resulting from a changing (time-varying) magnetic field is called an **induced emf**; any current caused by an induced emf is called an **induced current**. The process by which induced emfs and currents are produced is called **magnetic induction**.

Because magnetic forces act at right angles to the velocity of the particles they act on, they never do work. Therefore, we know that an induced emf is not the result of work done by a magnetic field. If the magnetic field is not doing the work, what is? The answer is that a changing magnetic field generates a *nonconservative* electric field, and it is this nonconservative electric field that circulates and, hence, does work on the charge carriers, which results in the induced emf. (Any vector field $\vec{E}$ for which $\oint_C \vec{E} \cdot d\vec{\ell} \neq 0$ is said to circulate.) The equation relating the rate of change of the magnetic field with the induced emf is known as Faraday's law:
$\mathcal{E} = \oint_C \vec{E} \cdot d\vec{\ell} = -\dfrac{d}{dt} \int_S \vec{B} \cdot \hat{n}\, dA = -\dfrac{d\phi_m}{dt}$ where ϕ_m is the magnetic flux through any surface bounded by the curve C.

The physical meaning of the negative sign in Faraday's law is referred to as Lenz's law, and is discussed in the next section.

Fundamental Equations

Faraday's law
$$\mathcal{E} = \oint_C \vec{E} \cdot d\vec{\ell} = -\frac{d}{dt} \int_S \vec{B} \cdot \hat{n}\, dA = -\frac{d\phi_m}{dt}$$

Common Pitfalls

> A large magnetic flux does not necessarily cause a large induced emf. It is the *rate* at which the flux changes, *not* the magnitude of the flux, that determines the induced emf. Thus, a large induced emf is caused by a rapidly changing magnetic flux.

> Steady magnetic fields do not induce an emf by themselves. If a coil rotates, there will be a changing flux through it, which will induce an emf.

1. TRUE or FALSE: Any change in the magnetic flux through a circuit results in an induced emf.

2. In hospitals with magnetic resonance imaging facilities and at other locations where large magnetic fields are present, there are usually signs warning people with pacemakers and other electronic medical devices not to enter. Why?

Try It Yourself #2

A 30-turn coil with a diameter of 6.00 cm is placed in a constant, uniform magnetic field of 1.00 T directed perpendicular to the plane of the coil. Beginning at time $t = 0$ the field is increased at a uniform rate until it reaches 1.30 T at $t = 10.0$ s. The field remains constant thereafter. What is the magnitude of the induced emf in the coil at (a) $t < 0$ s, (b) $t = 5.00$ s, and (c) $t > 10.0$ s?

Picture: Use Faraday's law to find the induced emf. The flux through a multiturn coil is equal to the flux per turn multiplied by the number of turns.

Solve:

Draw a sketch of the situation.	
Write Faraday's law as a guiding principle for the problem. In this problem neither the area nor the relative direction of the unit normal to the surface and the magnetic field is changing. Use these facts to simplify your expression *algebraically* for the induced emf.	
For $t < 0$ the magnetic field, and hence the magnetic flux through the coil, are not changing. Use this to find the induced emf for this timeframe.	$\mathcal{E} = 0$ for $t < 0$ s
At $t = 5.00$ s the magnetic field, and hence the flux through the coil, are changing. Find the time rate of change of the magnetic field.	
Now calculate the induced emf in the coil at $t = 5.00$ s.	$\mathcal{E} = 2.55$ mV for $t = 5.00$ s
For $t > 10.0$ s the rate of change of the magnetic field is again zero. Use this to find the induced emf for this timeframe.	$\mathcal{E} = 0$ for $t > 10.0$ s

Check: You should have a nonzero induced emf only when the flux is changing.

Taking It Further: Plot the magnetic field and the induced emf as functions of time for the range −5.00 s < t < 15.0 s.

Try It Yourself #3

A 50-turn square coil with a cross-sectional area of 5.00 cm² has a resistance of 20.0 Ω. The plane of the coil is perpendicular to a uniform magnetic field of 1.00 T. The coil is suddenly rotated about the axis shown in the figure through an angle of 60° over a period of 0.200 s. What charge flows past a point in the coil during this time?

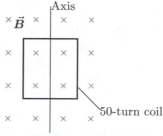

$\vec{B}$ into the page

Picture: Because the direction of the area vector changes relative to the magnetic field, the flux through the coil changes. This change in flux induces an emf in the coil, which causes current to flow in the coil. The total charge passing through one point is the integral of current over time.

Solve:

Obtain an *algebraic* expression for the total charge passing a single point. We know that current is voltage divided by resistance and that the voltage is really the induced emf in the loop. Note that the total charge that flows is ultimately independent of the speed at which the loop is rotated. Your final expression should include only the change in magnetic flux and the resistance.	
Determine the initial flux through the coil *algebraically*.	
Determine the final flux through the coil. After the coil has rotated 60° the magnetic field makes this same angle with the unit normal to the surface.	

Substitute the results of the second and third steps into the expression for the charge and solve.	
	$Q = 6.25 \times 10^{-4}$ C

Check: This is a reasonable amount of charge, and the units work out correctly.

Taking It Further: If the loop is rotated a full 360° around the axis, how much total charge passes the point in the loop? Explain.

28.3   Lenz's Law

In a Nutshell

The physical meaning of the negative sign in Faraday's law is so important it is known as Lenz's law. **Lenz's law** states, "The induced emf is in such a direction as to oppose, or tend to oppose, the change that produces it."

An alternative statement of Lenz's law in terms of the magnetic flux is also often useful: "When a magnetic flux through a surface changes, the magnetic field due to any induced current produces a flux of its own—through the same surface and opposite in sign to the initial change in flux."

When current flows through a long, tightly wound solenoid, a uniform magnetic field $\vec{B}$ is produced inside the solenoid. The magnetic flux per turn of the solenoid is BA, where A is the area of a plane surface bounded by a turn. The net magnetic flux through the solenoid is the flux per turn times the number of turns. When the current increases, the magnetic field increases, and so does the flux. In accordance with Lenz's law, an induced emf is directed so as to oppose the increasing flux due to the increasing current, and therefore is called a **back emf**.

Common Pitfalls

> Use Faraday's law to determine the magnitude of the induced emf. The minus sign in the equation is a useful reminder of Lenz's law, from which you can deduce the sense of the induced emf. It is always directed so that any induced current would oppose the change in flux.

> Lenz's law says that the induced emf is in a direction that opposes not the flux itself, but the *change* in the flux that produced it. The induced emf may be in a direction that tends to increase or decrease the existing flux, depending on the circumstances.

3. TRUE or FALSE: Lenz's law states that the direction of an induced emf is always opposite to the magnetic field that induced it.

4. Two conducting loops with a common axis are placed near each other, as shown, and initially the currents in both loops are zero. If a current is suddenly set up in loop a, as shown, is there also a current in loop b? If so, in which direction? What is the direction of the force that loop a exerts on loop b? Explain.

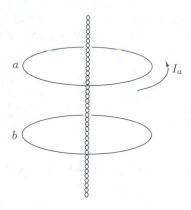

Try It Yourself #4

As shown in the figure, a square, 30-turn, 0.82-Ω coil 10.0 cm on a side is between the poles of a large electromagnet that produces a constant, uniform magnetic field of 6.00 kG. As suggested by the figure, the field drops sharply to zero at the edges of the magnet. The coil moves to the right at a constant velocity of 2.00 cm/s. What is the current through the coil wire (a) before the coil reaches the edge of the field, (b) while the coil is leaving the field, and (c) after the coil leaves the field? (d) What is the total charge that flows past a given point in the coil as it leaves the field?

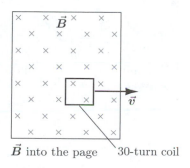

$\vec{B}$ into the page   30-turn coil

Picture: Faraday's law gives the induced emf in the coil, which will produce a flow of current in the coil. The flux through the coil changes as it moves out of the region of magnetic field.

Solve:

Determine the change in flux and hence the induced emf and current through the loop before it reaches the edge of the field.	
	$I = 0$ A
Determine the change in flux and hence the induced emf and current through the loop after the coil leaves the field.	
	$I = 0$ A
Calculate the rate of change of flux when the coil is partially out of the magnetic field by finding an expression for the flux when the right edge of the coil is a distance x from the edge of the field. A sketch will help.	

Relate the velocity of the coil to the rate of change of x and solve for the induced emf.	
Determine the current that flows through the coil while it is partially out of the magnetic field.	$I = 0.0439$ A
Determine the time required for the coil to leave the magnetic field.	
Integrate the current over this time interval to find the total charge that flows past a given point.	$Q = 0.220$ C

Check: These are reasonable values, and the units all work out.

Taking It Further: Plot the induced current in the loop as a function of the horizontal position of the right side of the current loop. Let the right-hand edge of the magnetic field region be $x = 0$. Your plot should be in the range of -5.00 cm $< x < 20.0$ cm.

Try It Yourself #5

Determine the magnitude and direction of the force on each side of the coil in the previous problem for situations (a) through (c).

Picture: We know the current in each case. Use this and the expression for the magnetic force on a wire to answer the question.

Solve:

The current for (a) and (c) is zero. Use this to determine the force on each segment of the coil.	$\vec{F} = 0$ on all segments.
Determine the direction of the current in the coil as it is leaving the magnetic field region.	
Using the expression for the magnetic force on a current-carrying wire, determine the force on the right-hand segment of the coil as it is leaving.	$\vec{F} = 0$
Do the same for the top segment of the loop. Your answer will have to be *algebraic*.	$\vec{F} = NIxB$, upward, where x is the length of the upper segment still in the region of nonzero magnetic field.
Do the same for the bottom segment of the loop. Your answer will have to be *algebraic*.	$\vec{F} = NIxB$, downward, where x is the length of the upper segment still in the region of nonzero magnetic field.

Do the same for the left-hand segment of the loop.	
	$\vec{F} = 0.0790$ N, to the left.

Check: According to Lenz's law, the induced current should oppose the change in magnetic flux. A net leftward force, which is what we found, satisfies that criterion.

Taking It Further: As the loop *enters* the field region from the left, what is the direction of the induced current and the resulting force on each segment of the coil?

28.4   Motional EMF

In a Nutshell

Motional emf is any emf induced by the motion of a conductor in a region in which there exists a magnetic field. The motional emf $\mathcal{E}$ is related to the magnetic field and the motion of the wire by $\mathcal{E} = -d\phi_m/dt$, where $d\phi_m$ is the magnetic flux through the surface that is swept out by the moving wire during time dt.

If a wire of length ℓ moves with velocity $\vec{v}$ through, and perpendicularly to, a magnetic field $\vec{B}$ the area dA swept out by the wire during time dt is $\ell v\, dt$. Thus $\mathcal{E} = -B\ell v$, where the minus sign is a reminder that the feedback is negative—that is, any currents generated by the motion of the wire are in such a direction that the magnetic force acting on the moving wire is directed opposite to its motion. Obviously, this force opposes the motion of the wire. This force will not be long-lived, since a steady current cannot be maintained in an isolated piece of wire.

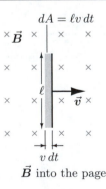

A **generator** converts mechanical energy into electrical energy, usually by moving a coil or coils of wire in a magnetic field, changing the flux through the coil, and generating an emf.

A **motor** is the inverse of a generator. By applying current to a current loop in a magnetic field, a torque is exerted on the coil, causing it to rotate. The result is a conversion of electrical energy to mechanical energy.

Important Derived Results

Magnitude of emf for a rod moving perpendicular to both the length of the rod and $\vec{B}$ $\mathcal{E} = B\ell v$

Common Pitfalls

> ➢ The fact that the emf from a generator varies sinusoidally with time has nothing to do with the shape of the rotating coil—only with the coil's rotation at constant angular speed in a static, uniform magnetic field.

5. TRUE or FALSE: Work is required to keep a sliding wire connected to an otherwise stationary circuit moving at a constant speed in the presence of a magnetic field.

6. A conducting rod slides without friction on conducting rails as shown. It is given an initial velocity $\vec{v}$ as shown. Describe its subsequent motion and justify your answer.

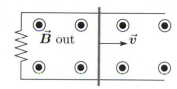

Try It Yourself #6

In 1996, the crew of the space shuttle performed a space tether experiment in which a 20.0-km-long electrical tether was strung out by the shuttle as it orbited Earth. As the tether was swept through Earth's magnetic field, a potential difference of approximately 3600 V was generated along the length of the tether. Assuming an approximately 400-km-high orbital radius, what was the magnitude of the magnetic field experienced by the tether?

Picture: We can find the speed of the shuttle, and hence of the tether, by applying the law of gravitation at the given orbital height. We can then use principle of motional emf to determine the magnetic field required to generate the given potential.

Solve:

Apply Newton's law of gravitation to the shuttle, assuming a circular orbit. Use this to solve *algebraically* for the speed of the shuttle and tether.	
Use the expression for the magnitude of emf for a rod moving perpendicular to a magnetic field, and solve *algebraically* for B.	
Substitute your result from the first step into the above expression.	

Finally, insert values, with units into your expression to solve for B.	
	$B = 0.235$ G

Check: We know Earth's magnetic field near the surface of Earth is approximately 1 G, so this seems reasonable.

Taking It Further: An actual current flows along the tether from the shuttle to a satellite at the end, and back to the shuttle through the ionosphere. As a result of this current, what will happen to the orbit of the shuttle?

Try It Yourself #7

The Faraday disk dynamo is a conducting disk rotated about its axis while in a magnetic field directed perpendicularly to the surface of the disk as shown. As the disk rotates, a radial line of the disk sweeps out an area, and an emf is induced between the center of the disk and its outside edge. When connected by a circuit, charges will flow. If the disk has radius R_{disk}, the magnitude of the field is B, the resistance is R, and the disk is rotated at an angular speed ω, find an algebraic expression for the current induced in the circuit.

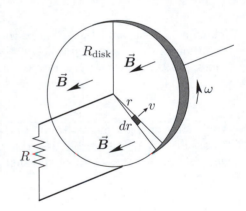

Picture: Different segments of the radial line are traveling at different speeds v, and so sweep out different areas in a given time. So we will need to consider the differential amount of emf $d\mathcal{E}$ induced along each differential length dr of the radial line.

Solve:

Consider a small segment of length dr of the radial line, a distance r from the axis. Determine the linear speed of this line segment. You may need to review rotational kinematics.	

From the speed above and the length of the differential segment, find an expression for the differential rectangular area swept out by the small radial line segment in some time dt.	
Using the area above, find an expression for the differential magnetic flux through that area, assuming B is perpendicular to the disk's surface.	
Determine the induced emf along that one differential segment by dividing both sides of the above expression by dt.	
To find the total emf induced along the radial line, integrate the above expression from the inner to outer radius of the disk.	$\mathcal{E}_{\text{disk}} = \frac{1}{2} B \omega R^2$
Use this emf and the resistance to find an expression for the current in the circuit.	$I = \dfrac{B \omega R_{\text{disk}}^2}{2R}$

Check: The units of this expression all work out.

Taking It Further: Is the center of the disk or the outside edge at a higher potential in this case? Explain.

28.5   Eddy Currents

In a Nutshell

When induced currents occur in bulk conductors (not thin wires), the induced currents tend to swirl around inside the conductors. These swirling currents are called **eddy currents**. In accordance with Lenz's law, if an eddy current is the result of a motional emf, a magnetic force acts on the moving conductor opposing its motion.

28.6   Inductance

In a Nutshell

At all points in space, the magnetic field due to a current I in a circuit is proportional to the current. Thus the flux of this magnetic field through the circuit, due to the circuit itself, is also proportional to I. So we can write $\phi_{\mathrm{m}} = LI$, where the proportionality constant L is called the **self-inductance** of the circuit. The value of the self-inductance of a circuit depends on the geometrical size and shape of the circuit. The SI unit of inductance is the henry (H). From the definition of self-inductance we have 1 henry (H) = 1 Wb/A.

In principle, the self-inductance L of any circuit can be calculated by assuming a current I, solving for the resulting magnetic field $\vec{B}$, computing the flux ϕ_{m} due to that current, and then using the expression $L = \phi_{\mathrm{m}}/I$ to determine the inductance. For most circuits this calculation is very difficult. However, it is straightforward for a long, tightly wound solenoid, for which $L = \mu_0 n^2 A \ell$, where ℓ is the length of the solenoid, n is the number of turns per unit length, and A is the cross-sectional area. A dimensional analysis of this equation shows us that the permeability of free space μ_0 can be alternatively expressed in henrys per meter.

Applying Faraday's law to the changing current in a circuit gives us $\mathcal{E} = -L\,dI/dt$—that is, the magnitude of the self-induced emf is proportional to the rate of change of the current in the circuit. The negative sign tells us that when the current is increasing, the self-induced emf opposes the increase, and when the current is decreasing, it opposes the decrease—that is, the feedback is always negative.

A circuit element like a coil or solenoid that has a significant self-inductance is called an **inductor**. The potential difference across an inductor with an internal resistance r is given by $\Delta V = -L\,dI/dt - Ir$.

Symbol for an inductor

Two or more circuits that are near each other influence each other by means of their magnetic fields. Consider circuit 1 with current I_1 which produces magnetic field $\vec{B}_1$ and circuit 2 with current I_2 which produces magnetic field $\vec{B}_2$. The field $\vec{B}_1$ is proportional to I_1 and the field $\vec{B}_2$ is proportional to I_2. The net magnetic flux $\phi_{\mathrm{m}2}$ through circuit 2 is the sum of two parts, one proportional to I_1 and one proportional to I_2. Thus $\phi_{\mathrm{m}2} = L_2 I_2 + M_{12} I_1$, where L_2 is the self-inductance of circuit 2 and M_{12} is called the **mutual inductance** of the two circuits. The mutual inductance depends on the individual geometries of the two circuits and their positions and orientations relative to each other. The flux through circuit 1, similarly, is $\phi_{\mathrm{m}1} = L_1 I_1 + M_{21} I_2$. It can be shown that the two mutual inductances are always equal, so it is convenient to drop the subscripts—that is, for any two circuits with a fixed geometry relative to each other, $M_{21} = M_{12} = M$.

Physical Quantities and Their Units

Inductance L or M SI unit of henrys (H); $1\ \mathrm{H} = 1\ \mathrm{Wb/A} = 1\ \mathrm{T \cdot m^2/A}$

Important Derived Results

Definition of self-inductance	$L = \dfrac{\phi_{\mathrm{m}}}{I}$
Self-inductance of a solenoid	$L = \mu_0 n^2 A \ell$
Mutual inductance	$M_{12} = \dfrac{\phi_{\mathrm{m}12}}{I_1}$
Mutual inductance of two solenoids	$M = \mu_0 n_2 n_1 \ell \pi r_1^2$

Common Pitfalls

> ➤ Inductance (either self or mutual) is a geometric and material property, and as such depends only on the sizes, shapes, and relative positions of circuits, as well as the material between the circuits, not on the currents in them.
> ➤ All circuits have some self-inductance—it's a matter of degree. Further, circuits that contain coils of wire with ferromagnetic cores have a great deal of inductance compared to those that do not. Because of inductance, any change in current is accompanied by negative feedback. Consequently, the current in circuits with significant inductance cannot instantaneously change from one value to another, as it can in circuits with negligible inductance. Instead, it must change incrementally.

7. TRUE or FALSE: Two circuits are close to each other but not in physical contact; if the current in one changes, an emf is induced in the other.

8. Explain why the mutual inductance of two circuits does not depend on the current in either circuit.

Try It Yourself #8

A tightly wound solenoid 18.0 cm long with a 2.00-cm diameter is made of 1500 turns of #22 copper wire. It is surrounded by a 20-turn circular coil, 3.00 cm in diameter, that is coaxial with the solenoid, as shown. The circular coil is connected across a resistor of very high resistance. What is the magnitude of the induced emf in the coil when the current in the solenoid is changing at a rate of 100 A/s?

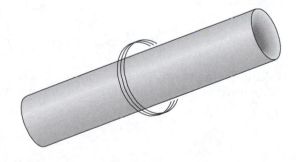

Picture: The emf induced in the coil is the emf induced by the change of the solenoid current plus the emf induced by the change of the circular-coil current.

Solve:

Write an *algebraic* expression for the induced emf in the coil in terms of mutual and self inductances. Make sure to use appropriate subscripts so you can keep straight the effects from the solenoid and the coil.	
The coil has a very high resistance, so it has very little current, so the self-inductance term is essentially zero. Simplify your expression above using this knowledge.	
Determine the flux through the coil from the solenoid. Make sure you take into account the physical extent of the magnetic field from the solenoid.	
The induced emf can now be found by substituting first expressions into your result, above, for the induced emf, and finally values with their units.	$|\mathcal{E}| = 6.58$ mV

Check: The units all work out, and this is a reasonable potential.

Taking It Further: How would you go about a reverse calculation? That is, how would you find the induced emf in the solenoid if the current in the coil were changing?

Try It Yourself #9

A Rowland ring, shown in the figure, can be used to study the properties of ferromagnetic materials. A toroidal coil is wound around a ring-shaped sample of the material, and an electrically separate pickup coil is wound around the toroid. In this case, the inner and outer radii of the toroid are both approximately 10.0 cm, and it is wound with 600 turns. The unmagnetized sample inside the toroid has a cross-sectional diameter of 0.800 cm, and the pickup coil has 50 turns with a resistance of 8.00 Ω. When the current in the toroid is increased from zero to 1.40 A, a total charge of 3.30×10^{-6} C flows in the pickup coil. Find B_{app}, M, and B in the coil.

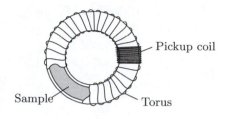

Picture: The total charge that flows through the pickup coil is due to the changing flux through the pickup coil.

Solve:

Relate *algebraically* the total flow of charge to the change in magnetic flux through the pickup coil.	
Find an *algebraic* expression for the total magnetic flux through the pickup coil, in terms of B.	
Use your expressions from the first and second steps to solve for B.	$B = 0.0105$ T
Find B_{app} from the expression for the magnetic field inside a tightly wound toroid derived in Chapter 27.	$B_{\text{app}} = 1.68 \times 10^{-3}$ T

Use your values from the third and fourth steps to find M.	
	$M = 7080$ A/m

Check: All these values are reasonable, and the units are all consistent.

Taking It Further: How much total charge would flow in the pickup coil if the ferromagnetic material were not present?

28.7 Magnetic Energy

In a Nutshell

Like a capacitor, an inductor stores energy. The energy U stored in an inductor is given by $U = \frac{1}{2}LI^2$. It is useful to consider the stored energy as being stored in the magnetic field. The **energy density of the magnetic field** is $u_\mathrm{m} = B^2/(2\mu_0)$.

Important Derived Results

Energy stored in an inductor $$U = \frac{1}{2}LI^2$$

Magnetic field energy density $$u_\mathrm{m} = \frac{B^2}{2\mu_0}$$

Common Pitfalls

> ➤ The magnetic and electric field energy densities are given by similar expressions. Do not confuse them.

9. TRUE or FALSE: No energy is required to sustain a constant current in an inductor with negligible resistance.

10. Compare the expressions for the energy stored in an inductor and the energy stored in a capacitor.

Try It Yourself #10

A tightly wound solenoid of 1600 turns, cross-sectional area of 6.00 cm^2, and length of 20.0 cm carries a current of 2.80 A. (a) What is its inductance? (b) How much energy is stored in the solenoid?

Picture: Apply the expressions that have been derived for these quantities.

Solve:

Find the inductance using the expression for the inductance of a long, tightly wound solenoid.	
	$L = 9.65 \times 10^{-3}$ H
Find the energy stored in the inductor by using the appropriate expression.	
	$U = 0.0378$ J

Check: The units all work out properly.

Taking It Further: If the cross-sectional area is doubled, what, if anything, happens to the values of L and U? Explain.

28.8   *RL* Circuits

In a Nutshell

Applying Kirchhoff's law to the circuit at right yields the following differential equation for an energizing *RL* circuit once the switch is closed: $\mathcal{E} - IR - L\dfrac{dI}{dt} = 0$. Integrating this equation we obtain an expression for the current in the circuit as a function of time after the switch is closed at $t = 0$: $I = \dfrac{\mathcal{E}}{R}\left(1 - e^{-Rt/L}\right) = I_f\left(1 - e^{-t/\tau}\right)$, where $\tau = L/R$ is the **time constant** of the circuit and I_f is the current after the circuit has been connected for a long time.

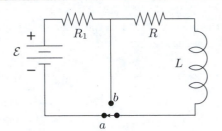

Similarly, when an inductor with a current in it is simultaneously disconnected from a battery and connected across a resistor, as would happen when the switch is moved from position a to position b, applying Kirchhoff's loop rule results in the expression $-IR - L\dfrac{dI}{dt} = 0$. Integrating this expression we find the current as a function of time after the switch is moved to position b at $t = 0$ is given by $I = I_0\, e^{-Rt/L} = I_0\, e^{-t/\tau}$, where I_0 is the initial current.

Important Derived Results

Current in an energizing *RL* circuit $I = \dfrac{\mathcal{E}}{R}\left(1 - e^{-Rt/L}\right) = I_f\left(1 - e^{-t/\tau}\right)$

Current in a de-energizing *RL* circuit $I = I_0\, e^{-Rt/L} = I_0\, e^{-t/\tau}$

RL time constant $\tau = \dfrac{L}{R}$

Common Pitfalls

> Remember that an energizing *RL* circuit starts with no current, and as $t \to \infty$ the current approaches a steady-state value. A de-energizing *RL* circuit starts with a finite initial current that decays to zero as $t \to \infty$.

11. TRUE or FALSE: If an ideal battery is connected across an inductor that has negligible resistance, the current rises to its steady-state value with a very short time constant.

12. When the switch S is opened in the RL circuit shown, a spark jumps between the switch contacts. Why?

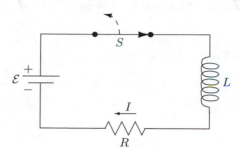

Try It Yourself #11

A tightly wound solenoid 18.0 cm long with a 2.00-cm diameter is made of 1500 turns of #22 copper wire. Find the (a) inductance, (b) resistance, and (c) time constant of the solenoid.

Picture: The inductance can be found by using the expression for the inductance of a tightly wound solenoid. You will need the resistivity of copper and the cross-sectional area of #22 copper wire, found in Tables 25-1 and 25-2 on page 847 of the text.

Solve:

Find the inductance using the expression for the inductance of a long, tightly wound solenoid.	$L = 4.94 \times 10^{-3}$ H
Determine the length of wire needed to build this solenoid.	
From the length of wire and the information from Tables 25-1 and 25-2, on page 847 of the text, determine the resistance of the wire.	$R = 4.92 \ \Omega$
The time constant can be determined from the inductance and the resistance.	$\tau = 1.00$ ms

Check: All the units work out properly.

Taking It Further: Use this space to do a careful units analysis and convince yourself that the units of H/ohm are in fact seconds.

Try It Yourself #12

The switch shown is closed and the current through the resistor at some instant is 6.00 A to the left. What is the rate of change of the current at this instant? Is the current increasing or decreasing? What is the potential difference $V_a - V_b$?

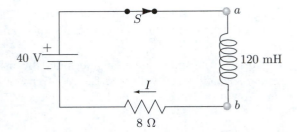

Picture: Apply Kirchhoff's loop rule to the circuit and solve for the potential difference across the inductor. Use this to find the rate of change of the current.

Solve:

Apply Kirchhoff's loop rule to find the potential difference across the inductor.	$V_a - V_b = -8.00$ V
Use $V_b - V_a$ to determine if the current is increasing or decreasing.	Current is decreasing.
Determine the rate of change of the current using the expression for the voltage gain or drop across an inductor.	$\dfrac{dI}{dt} = -66.7$ A

Check: All the units work out properly.

Taking It Further: What would have to be true about the current for us to deduce that it is increasing?

28.9 Magnetic Properties of Superconductors

In a Nutshell

As a superconductor is cooled below the critical temperature in an applied magnetic field, the magnetic field inside the superconductor becomes zero. This is known as the **Meissner effect**. Superconducting currents on the surface of the superconductor are induced that produce a magnetic field in the opposite direction of the applied field that exactly cancels the applied field—that is, a superconductor is a perfect diamagnetic material.

One of the results of a quantum-mechanical treatment of superconductivity is that the magnetic flux is **quantized**, and only appears in increments of $h/(2e)$, a quantity called a **fluxon**.

Important Derived Results

Quantization of magnetic flux
$$\phi_\mathrm{m} = n\frac{h}{2e}$$

QUIZ

1. TRUE or FALSE: Inductance is a geometric property.

2. TRUE or FALSE: A current cannot change abruptly in a circuit that has an inductance.

3. A common physics demonstration is to drop a bar magnet down a long, vertical aluminum pipe. Describe the motion of the bar magnet and the physical explanation for it.

4. An electric field is set up in the conducting bar as the bar moves to the right as shown. What causes this field and in what direction does it point? Does an external force have to act on the bar to keep it moving at a constant speed?

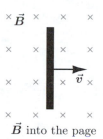

$\vec{B}$ into the page

5. Is the electric field created by a changing magnetic flux conservative or nonconservative? Explain.

6. After the switch in the figure is closed, charge begins to flow. At the instant at which half the power supplied by the battery is being dissipated in the resistor, what is the current and how fast is it changing?

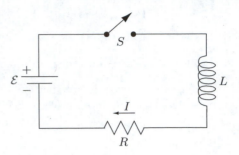

7. A 20-turn circular coil, 1.60-cm in diameter, is inside and coaxial with a tightly wound solenoid 18.0 cm long with a 2.00-cm diameter made of 1500 turns of #22 copper wire as shown. The circular coil is connected across a resistor of very high resistance. What is the magnitude of the emf induced in the circular coil when the current in the solenoid changes at a rate of 100 A/s?

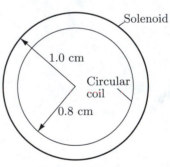

End-on view

Chapter 29

Alternating-Current Circuits

29.1 Alternating Current in a Resistor

In a Nutshell

A simple **ac generator** uses a changing magnetic flux or a motional emf to generate a sinusoidal emf given by $\mathcal{E} = \mathcal{E}_{\text{peak}} \cos \omega t$. If generated by a coil with N turns and area A in a magnetic field B, the peak emf is given by $\mathcal{E}_{\text{peak}} = \omega N B A$.

Symbol for an ac generator

The circuit shown contains an ideal ac generator with no internal resistance, capacitance, or inductance, and a resistor. Application of Kirchhoff's law reveals that the voltage across the resistor is equal to the emf of the generator: $\mathcal{E} = V_R = V_{R\,\text{peak}} \cos \omega t$. Both the peak and time-varying potential and current obey the same relationship as Ohm's law for dc circuits, as long as we look at the peak or rms (see below) values of both quantities. The general expression for power is also the same, although the power now also varies with time: $P = I^2 R = I^2_{\text{peak}} R \cos^2 \omega t$. The **average power** is given by $P_{\text{av}} = \frac{1}{2} I^2_{\text{peak}} R$ because the average value of $\cos^2 \omega t$ is $\frac{1}{2}$.

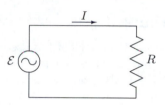

Often in ac circuits the **root-mean-square** (rms) of a time-varying function is used rather than the peak value. This value is useful because the rms current, for instance, equals the steady dc current that would produce the same Joule heating as the actual ac current. The rms value of a function is defined as $I_{\text{rms}} = \sqrt{(I^2)_{\text{av}}}$—that is, it is the square root of the average of the square of the function. For all sinusoidally varying functions, the rms value is equal to the peak value of the function divided by $\sqrt{2}$: for example, $I_{\text{rms}} = I_{\text{peak}}/\sqrt{2}$. Ohm's law also holds for rms values of the current and voltage.

Applying the definitions for rms current and emf to the expression for average power, it can be shown that the average power delivered by a generator is given by $P_{\text{av}} = \mathcal{E}_{\text{rms}} I_{\text{rms}}$. This expression also works for the power dissipated in a resistor if the emf is replaced by the voltage across the resistor.

Domestic power in the United States has an rms potential difference of approximately 120 V at a frequency of 60 Hz. All electrical devices in the same circuit are typically attached in parallel. To prevent excess current draw that might melt the wiring, each circuit typically has a 15-A or 20-A circuit breaker or fuse that trips or blows when the specified current has been exceeded. Circuits with higher voltage and current ratings are also used, especially for appliances like stoves, dryers, and air conditioners.

Important Derived Results

Ideal generator emf $\mathcal{E} = \mathcal{E}_{peak} \cos \omega t = \omega NBA \cos \omega t$

Definition of root-mean-square (rms) $I_{rms} = \sqrt{(I^2)_{av}} = \dfrac{I_{peak}}{\sqrt{2}}$

Average power dissipated by a resistor $P_{av} = \dfrac{1}{2} I_{peak}^2 R = I_{rms}^2 R$

Common Pitfalls

> The average value of the square of any sinusoidal function over one or more complete cycles is $\frac{1}{2}$. Also, for any sinusoidally varying function, the maximum value, or amplitude, equals $\sqrt{2}$ times the rms value.

> Remember the conversion factor between ω and f: $\omega = 2\pi f$.

1. TRUE or FALSE: The emf of an ac generator varies sinusoidally with time.

Try It Yourself #1

An 8.00×10^{-2}-Ω power cord is used to deliver 1500 W of power. (a) If the power is delivered at 12.0 V rms, how much power is dissipated in the power cord (assuming the current and voltage are in phase)? (b) If the power is delivered at 120 V rms, how much power is dissipated in the power cord (again assuming the current and voltage are in phase)?

Picture: Not all the power will be dissipated in the power cord. Some will be dissipated by the device it is powering.

Solve:

Find the current required to deliver 1500 W of power at 12 V rms.	
Determine the average power dissipated by the cord when this much current flows through it.	$P_{av} = 1250$ W
Find the current required to deliver 1500 W of power at 120 V rms.	

Determine the average power dissipated by the cord when this much current flows through it.	
	$P_{\text{av}} = 12.5$ W

Check: The units work out, and in both cases the power dissipated by the cord is less than the total power delivered.

Taking It Further: Which voltage would you prefer to use to power your electrical device? Why?

29.2   Alternating-Current Circuits

In a Nutshell

Consider the simple circuit shown. The voltage across the inductor is equal to the emf of the generator, so we have $\mathcal{E} = V_L = \mathcal{E}_{\text{peak}} \cos\omega t = L\, dI/dt$ for the current direction given. Solving for the current, we find that $I = I_{\text{peak}} \sin\omega t$. Because of the phase difference between sine and cosine, we say that the potential drop across an inductor *leads the current* by 90°— that is, the maximum voltage across the inductor occurs before the maximum current through the inductor. 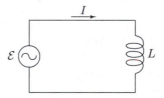

Further analysis of the current and emf leads us to define the **inductive reactance** $X_L = \omega L$. Like resistance, reactance has units of ohms, and an expression similar to Ohm's law relating the voltage drop and current holds for reactance: $I_{\text{peak}} = V_{L\ \text{peak}}/X_L$ and $I_{\text{rms}} = V_{L\ \text{rms}}/X_L$.

The instantaneous power delivered to an inductor is, as always, the product of the current through and voltage drop across the inductor. However, because these vary sinusoidally, 90° out of phase, the resulting expression is $P = \frac{1}{2} I_{\text{peak}} V_{L\ \text{peak}} \sin(2\omega t)$. The average power delivered to an ideal inductor over one complete cycle is therefore zero.

The voltage across the capacitor in the simple circuit shown is equal to the emf of the generator, so we have $\mathcal{E} = V_C = \mathcal{E}_{\text{peak}} \cos\omega t = Q/C$. Differentiating to find the current, we have $I = -I_{\text{peak}} \sin\omega t$, and the current is again 90° out of phase with respect to the potential across the capacitor. However, for capacitors, the maximum voltage occurs *after* the maximum in the current, so we say that the potential drop across a capacitor *lags the current* by 90°. 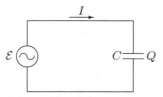

The **capacitive reactance** is given by $X_C = 1/(\omega C)$, again with units of ohms. The same relationship as for inductors holds for current through and voltage across a capacitor: $I_{rms} = V_{C\ rms}/X_C$ and $I_{peak} = V_{C\ peak}/X_C$.

The power delivered to a capacitor is also still the product of current and voltage, yielding $P = \frac{1}{2}I_{peak}V_{C\ peak}\sin(2\omega t)$, the same as for inductors.

Physical Quantities and Their Units

Reactance X SI units of ohms (Ω)

Important Derived Results

Inductive reactance $X_L = \omega L$

rms current through an inductor $I_{rms} = \dfrac{V_{L\ rms}}{\omega L} = \dfrac{V_{L\ rms}}{X_L}$; voltage leads current by $90°$

Capacitive reactance $X_C = \dfrac{1}{\omega C}$

rms current "through" a capacitor $I_{rms} = \omega C V_{C\ rms} = \dfrac{V_{C\ rms}}{X_C}$; voltage lags current by $90°$

Common Pitfalls

> ➤ In an ac circuit the instantaneous current through a device is not necessarily in phase with the instantaneous voltage drop across it.
> ➤ A reactance is not a resistance, although both are measured in ohms. There is no power dissipation in a purely reactive device for which the voltage drop and current are $90°$ out of phase.
> ➤ Capacitive and inductive reactance have opposite frequency dependencies. A capacitor has a very large reactance at low frequencies and a very small reactance at high frequencies; the reverse is true for an inductor.

2. TRUE or FALSE: The voltage drop across a capacitor lags behind the current by $90°$.

3. What is meant by this statement: "The voltage drop across an inductor leads the current by $90°$"?

Try It Yourself #2

A sinusoidal voltage of 40.0 V rms and a frequency of 100 Hz is applied to (a) a 100-Ω resistor, (b) a 0.200-H inductor, and (c) a 50.0-μF capacitor. Find the peak value of the current in each case.

Picture: Use the resistance or reactance for the appropriate circuit element. Convert the rms current to peak current.

Solve:

Write an *algebraic* expression relating the rms current, voltage, and resistance.	
Convert the rms current to a peak current and solve.	$I_{R \text{ peak}} = 0.566$ A
Write an *algebraic* expression relating the rms current, voltage, and inductive reactance.	
Convert the rms current to a peak current and solve.	$I_{L \text{ peak}} = 0.449$ A
Write an *algebraic* expression relating the rms current, voltage, and capacitive reactance.	
Convert the rms current to a peak current and solve.	$I_{C \text{ peak}} = 1.78$ A

Check: These units all work out properly, and the values are reasonable.

Taking It Further: What is the average power delivered in each case? Explain.

Try It Yourself #3

A sinusoidal voltage of 50.0 V (peak) at a frequency of 400 Hz is applied to a capacitor of unknown capacitance. The current in the circuit is 400 mA rms. What is the capacitance?

Picture: Determine the reactance of the circuit and use that to solve for the capacitance.

Solve:

Convert the peak voltage to rms voltage.	
Use the rms current and voltage to compute the reactance of the circuit.	
Use the reactance to calculate the capacitance.	$C = 4.50 \ \mu\text{F}$

Check: The units are correct, and the value is reasonable.

Taking It Further: If the frequency of the voltage is increased, what, if anything, will happen to the rms value of the current in the circuit? Why?

29.3   The Transformer*

In a Nutshell

A step-up transformer is a device that increases voltage at the expense of current. A step-down transformer does the opposite. An ideal transformer consumes no energy, so the power delivered to the primary circuit equals the power delivered by the secondary circuit. Because the two coils have the same magnetic flux per turn through them, the induced emf in each coil is proportional to the number of its turns. In an ideal transformer, $\frac{V_{1\ \text{rms}}}{V_{2\ \text{rms}}} = \frac{I_{2\ \text{rms}}}{I_{1\ \text{rms}}} = \frac{N_1}{N_2}$, where N_1 and N_2 are the numbers of turns in the primary and secondary coils, respectively. Our ability to step voltages up and down makes long-distance power transmission, at high voltage and low current, with acceptable energy losses possible.

Important Derived Results

Relations for an ideal transformer
$$\frac{V_{1\ \text{rms}}}{V_{2\ \text{rms}}} = \frac{I_{2\ \text{rms}}}{I_{1\ \text{rms}}} = \frac{N_1}{N_2}$$

Common Pitfalls

> The number of turns does not determine if a coil is the primary or secondary. An ac voltage *source* is always applied to the primary, and the secondary coil drives an electrical *load*.

4. TRUE or FALSE: Transformers are designed to maximize the mutual inductance of the primary and secondary coils.

5. Discuss this statement: "A transformer whose primary coil has 10 times as many turns as its secondary coil normally delivers about 10 times more power than it receives."

Try It Yourself #4

The 400-turn primary coil of a step-down transformer is connected to a 110-V rms ac line. The secondary coil is to supply 15.0 A at 6.30 V rms. Assuming no power loss in the transformer, find (a) the number of turns in the secondary coil and (b) the current in the primary coil.

Picture: The voltage ratio equals the ratio of the number of turns of the coils. The input power must be the same as the output power.

_____

*Optional material

Solve:

Determine the number of turns in the secondary coil using the ratio of voltages and number of turns. A fractional number of turns is not practical.	$N_2 = 23$ turns
Find the required input current by using the fact that the output power must equal the input power since there are no power losses.	$I_{\text{in rms}} = 0.859$ A

Check: These values are all reasonable, and the units work out properly.

Taking It Further: Use this space to calculate the power in both the primary and secondary coils to confirm that there is no energy loss.

29.4   *LC* and *RLC* Circuits without a Generator*

In a Nutshell

Applying Kirchhoff's loop rule to a circuit consisting of an initially charged capacitor and an inductor such as shown in the figure gives us an equation that is identical to the equation for simple harmonic motion. The solution to this equation is that the charge on the capacitor and the current vary as $Q = Q_{\text{peak}} \cos(\omega t)$ and $I = -I_{\text{peak}} \sin(\omega t)$, where $\omega = 1/\sqrt{LC}$. When the magnitude of the charge on the capacitor is at a maximum, the current is zero, and all the energy is stored in the capacitor as electrical potential energy. When the magnitude of the current is at a maximum, the stored charge is zero, and all the energy is stored in the inductor as magnetic potential energy. If a resistor is included in our circuit, we obtain an equation identical to that of a linearly damped harmonic oscillator.

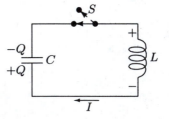

*Optional material

Important Derived Results

Current in an LC circuit without a generator $\qquad I = -I_{\text{peak}} \sin(\omega t)$

Charge on a capacitor in an LC circuit without a generator $\qquad Q = Q_{\text{peak}} \cos(\omega t)$

Common Pitfalls

6. Discuss the storage of energy in an ideal LC circuit with no losses.

29.5 Phasors*

In a Nutshell

For a resistor the voltage drop is in phase with the current; for an inductor the voltage drop leads the current by a phase angle of 90°; and for a capacitor the voltage drop lags behind the current by a phase angle of 90°. If the three components are connected in series as shown then the current through each of them is $I = I_{\text{peak}} \cos \theta$, where $\theta = \omega t - \delta$ is the phase of the current. The voltages across each element are out of phase, however:

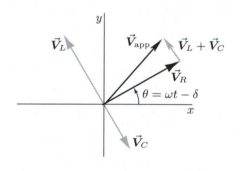

$$V_R = V_{R,\text{ peak}} \cos(\theta)$$
$$V_L = V_{L,\text{ peak}} \cos\left(\theta + \frac{\pi}{2}\right)$$
$$V_C = V_{C,\text{ peak}} \cos\left(\theta - \frac{\pi}{2}\right)$$

and the voltage drop V_{app} across the series combination is given simply by the sum of the voltages at any one instant in time.

With each voltage drop proportional to the cosine of an angle, you can find the x component of a resultant vector by adding the x components of the vectors. The two-dimensional vectors shown in the figure corresponding to the voltage drops are called **phasors** and have magnitudes equal to the maximum voltages $V_{R,\text{ peak}}$, $V_{L,\text{ peak}}$, and $V_{C,\text{ peak}}$. The three phasors rotate with the same angular velocity ω. This means that relative to one another their phases are fixed. To find the voltage drop across the series combination, you add the phasors (using vector addition of course) and then take the x component of the sum.

*Optional material

29.6   Driven RLC Circuit[*]

In a Nutshell

Applying Kirchhoff's loop rule to a circuit consisting of a generator connected to a series combination of an inductor, a capacitor, and a resistor, as shown, gives us the equation $L\dfrac{d^2Q}{dt^2} + R\dfrac{dQ}{dt} + \dfrac{1}{C}Q = \mathcal{E}\cos(\omega t)$, which has the steady-state solution $I = I_{\text{peak}}\cos(\omega t - \delta)$. Here the **phase angle** δ is given by $\tan\delta = (X_L - X_C)/R$ and the peak current is $I_{\text{peak}} = V_{\text{app peak}}/\sqrt{R^2 + (X_L - X_C)^2} = V_{\text{app peak}}/Z$. The quantity $X_L - X_C$ is called the **total reactance** of the circuit, and Z is the **impedance** of the circuit.

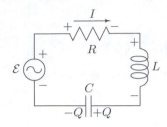

The peak current is largest when the total reactance, $X_L - X_C$, is zero. The expression for the frequency ω_{res} at which this occurs, called the **natural** or **resonance frequency**, can be computed by equating the expressions for the inductive and capacitive reactances and solving for the frequency. This exercise results in $\omega_{\text{res}} = 1/\sqrt{LC}$. When the total reactance is zero the phase angle δ is also zero; so, at resonance, the current is in phase with the generator voltage.

The average power delivered to a series RLC combination is all delivered to the resistor. This power is $P_{\text{av}} = V_{\text{app rms}}I_{\text{rms}}\cos\delta$, where $V_{\text{app rms}}$ and I_{rms} are, respectively, the rms voltage drop across and the rms current through the entire series–connected three-element combination. The quantity $\cos\delta$ is called the **power factor**. The power meters found on most homes do not measure the integrated power or energy but rather the integrated product $V_{\text{app rms}}I_{\text{rms}}$, which just happens to equal the energy when the power factor is 1.

Plots of P_{av} versus frequency are called **resonance curves**. The measure of the sharpness of the resonance is called the **Q factor**, which is given by $Q = \omega_{\text{res}}L/R \approx \omega_{\text{res}}/(\Delta\omega)$; $\Delta\omega$, the full width of the resonance curve at half the peak power, is called the **resonance width**.

When an inductor, a capacitor, and a resistor are connected in parallel across the terminals of a generator, it is the currents that add; but because they are not in phase, it is convenient to represent them as phasors and add them vectorally. The impedance is related to the resistance and the reactances by the equation $\dfrac{1}{Z} = \sqrt{\dfrac{1}{R^2} + \left(\dfrac{1}{X_L} - \dfrac{1}{X_C}\right)^2}$. The peak current is related to the amplitude of the voltage supplied by the generator by $I_{\text{peak}} = \mathcal{E}_{\text{peak}}/Z$.

[*]Optional material

Important Derived Results

Total reactance of a series RLC circuit	$X_L - X_C$
Definition of impedance	$Z = \dfrac{V_{rms}}{I_{rms}}$
Impedance of a series RLC circuit	$Z = \sqrt{R^2 + (X_L - X_C)^2}$
Current in a series RLC circuit	$I = I_{peak} \cos(\omega t - \delta)$
Phase angle	$\tan \delta = (X_L - X_C)/R$
Resonance frequency of a series RLC circuit	$\omega_{res} = 1/\sqrt{LC}$
Average power delivered to a series RLC circuit	$P_{av} = V_{app\ rms} I_{rms} \cos \delta$
Power factor	$\cos \delta$
Q factor	$Q = \dfrac{\omega_{res} L}{R} \approx \dfrac{\omega_{res}}{\Delta \omega}$
Impedance of a parallel RLC circuit	$\dfrac{1}{Z} = \sqrt{\dfrac{1}{R^2} + \left(\dfrac{1}{X_L} - \dfrac{1}{X_C}\right)^2}$

Common Pitfalls

> ➤ Because of phase differences, maximum or rms voltage drops in a series ac circuit cannot simply be added as if they were scalars. One way to add them is vectorally, using phasors.
> ➤ The behavior of a series RLC circuit near resonance depends on all three quantities, but the resonance frequency is determined by L and C independent of R.

7. TRUE or FALSE: The voltage drop across the resistor in a parallel RLC circuit is in phase with the voltage drop across the capacitor.

Try It Yourself #5

The figure shows a series RLC circuit with $L = 0.0100$ H, $R = 220$ Ω, and $C = 0.100$ μF. The amplitude of the driving voltage is $\mathcal{E}_{max} = 150$ V. At a frequency of 1000 Hz, what power is being supplied by the generator?

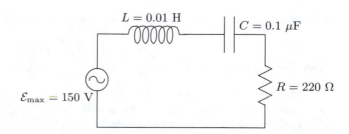

Picture: All the power supplied by the generator is dissipated by the resistor. The impedance of the circuit can be used to determine the current and hence the power supplied.

Solve:

Find the inductive reactance.	
Find the capacitive reactance.	
Calculate the impedance of the circuit.	
Calculate the maximum current through the circuit using the maximum emf and the impedance.	
Calculate the power delivered by the generator. All the power is dissipated by the resistor because the capacitor and inductor are nondissipative.	$P = 1.04$ W

Check: All the units are correct, and this is a reasonable value for the power.

Try It Yourself #6

In the series *RLC* circuit in the figure find the maximum voltages V_R, V_C, and V_L across the resistor, capacitor, and inductor, respectively, (a) at a frequency of 100 Hz and (b) at resonance.

Picture: The impedance of the circuit at each frequency can be used to determine the current and then the voltage across each element.

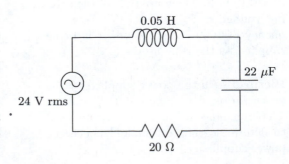

Solve:

Determine the resonance frequency of the circuit.	
Calculate the impedance at resonance.	
Calculate the impedance of the inductor at 100 Hz.	
Calculate the impedance of the capacitor at 100 Hz.	
Determine the total impedance at 100 Hz.	
Determine the maximum current at 100 Hz from the impedance and the maximum emf.	
Determine the maximum voltage across the resistor from the maximum current and the resistance.	$V_{R\ max} = 14.9$ V

Determine the maximum voltage across the inductor from its impedance and the maximum current.	
	$V_{L\,max} = 23.4$ V
Determine the maximum voltage across the capacitor from its impedance and the maximum current.	
	$V_{C\,max} = 53.9$ V
Determine the maximum current at resonance from the impedance and the maximum emf.	
Determine the maximum voltage across the resistor from the maximum current and the resistance.	
	$V_{R\,max} = 34.0$ V
Determine the maximum voltage across the inductor from its impedance and the maximum current.	
	$V_{L\,max} = 80.9$ V
Determine the maximum voltage across the capacitor from its impedance and the maximum current.	
	$V_{C\,max} = 80.9$ V

Check: All the units check out, and these seem like reasonable values.

Taking It Further: (a) Explain why the sum of the maximum voltages in each case is larger than the maximum voltage of the applied emf.   (b) Why is the maximum voltage across the inductor equal to the maximum voltage across the capacitor at resonance?

QUIZ

1. TRUE or FALSE: The voltage drop across the inductor in a series *RLC* combination is 180° out of phase with the voltage drop across the capacitor.

2. TRUE or FALSE: A capacitor acts like a short circuit at very high frequencies.

3. Why is electric power for domestic use in the United States transmitted at very high voltages and stepped down to 110 V by a transformer near the point of consumption?

4. In a coil that has both resistance and inductance, does the phase angle between the current through the coil and the voltage drop across it vary with the frequency?

5. Why is the total or equivalent reactance of a series *LC* combination the difference between the inductive and capacitive reactances rather than the sum?

6. A series RLC circuit is driven by 115 V rms from an ac power line of unknown frequency. The power drawn is 65.0 W and the current in the circuit is 1.00 A rms. Find (a) the resistance and (b) the net reactance of the circuit.

7. A series RLC circuit consists of an 8.00-Ω resistor, a 100-mH inductor, and a 5.00-μF capacitor. A signal generator applies 5.00 V rms to this circuit. (a) At what frequency is the maximum power delivered to this circuit? (b) What is this maximum power?

Chapter 30

Maxwell's Equations
and Electromagnetic Waves

30.1   Maxwell's Displacement Current

In a Nutshell

Ampère's law ($\oint_C \vec{B} \cdot d\vec{\ell} = \mu_0 I_S$) is valid only for steady-state situations—that is, situations in which all currents and all charge distributions are constant in time. The current that appears in Ampère's law is the rate of flow of charge through any surface S that is bounded by the closed curve C. Maxwell modified this relation to include situations in which the currents and charges are *not* constant. The resulting relation is $\oint_C \vec{B} \cdot d\vec{\ell} = \mu_0(I_S + I_d)$, where $I_d = \epsilon_0 \, d\phi_e/dt$ is called the **displacement current** and ϕ_e is the net electric flux out of the surface bounded by the curve C. The displacement current is not actually a current; that is, it is not a flow of charge through a surface. However, it contributes to the net magnetic field in the same way that a current does.

Fundamental Equations

Generalized form of Ampère's law

$$\oint_C \vec{B} \cdot d\vec{\ell} = \mu_0(I + I_d) = \mu_0 I + \mu_0 \epsilon_0 \frac{d\phi_e}{dt}$$

Important Derived Results

Displacement current

$$I_d = \epsilon_0 \frac{d\phi_e}{dt}$$

Common Pitfalls

> ➤ Remember: The original form of Ampère's law holds only when there are no time-varying electric fields. This occurs only when all currents and all charge distributions are constant. However, the modified Ampère's law holds even when electric fields vary with time.

1. TRUE or FALSE: The magnetic field produced by a displacement current has properties identical to those of the magnetic field produced by an ordinary current.

2. Why is the idea of displacement current necessary?

Try It Yourself #1

The parallel-plate capacitor shown in the figure has circular plates of radius 1.00 m separated by a 0.500-mm
thick sheet of mica. The dielectric constant of mica is 5.4.
Charge is flowing onto one plate and off the other plate at a
rate of 5.00 A. (a) Find the displacement current through
a 1.500-m-radius circular surface S, coaxial with the capacitor, that is in the gap between the plates. (b) Find
the current through S. (c) Find the magnetic field on the
perimeter of S.

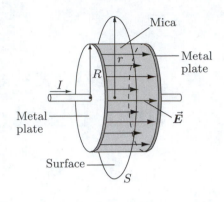

Picture: This example requires the use of Ampère's law
with the Maxwell correction. The changing electric flux
through the surface is the key element of this problem.
Figure (a) shows a side view of the capacitor with the
surface and bound charges and the resulting electric field.
Figure (b) shows a cross-sectional view of the capacitor,
the electric field, and the ampèrian loop.

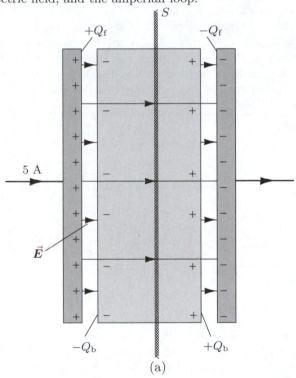

(a)

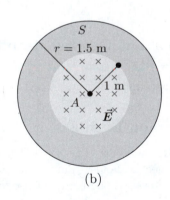

(b)

Solve:

Find an *algebraic* expression for the displace-ment current in terms of the area and the elec-tric field between the plates. The electric field is everywhere parallel to the normal of the sur-face vector. The electric field is also uniform. Because the area is constant, we need to worry only about changes in the electric field causing a change in the electric flux.	

Determine an *algebraic* expression for the electric field that gives rise to the flux through the surface. This expression should be in terms of the charge on the plates, an area, and constants. The electric field produced by the free charge is reduced by the dielectric.	
Substitute this electric field into the expression for the displacement current and solve.	$I_d = 0.926$ A
Solve for the actual physical current that flows through the surface. This current is due to the polarization of the bound charges in the mica. As discussed on page 827 in Chapter 24 of the text, the bound charge is related to the free charge.	$I = 4.07$ A
Apply Ampère's law *algebraically* with the Maxwell correction to find the magnetic field along surface S. Because of the symmetry in the problem, the magnetic field is everywhere parallel to $d\vec{\ell}$ and is also uniform.	
Relate the polarization and displacement currents to the free charge.	

Substitute values and units to solve for the magnetic field.	
	$B = 6.67 \times 10^{-7}$ T

Check: These are reasonable values, and the units work out.

Taking It Further: How would this problem change if, instead of mica, the capacitor were filled with vacuum?

Try It Yourself #2

For the situation described in Try It Yourself #1, find the magnetic field in the mica at a distance of 30.0 cm from the axis of the capacitor. The magnetic susceptibility χ_m of mica is negligible.

Picture: This problem will be solved in the same way we solved Try It Yourself #1. However, not all of the polarization current will contribute to the flux through the surface.

Solve:

Sketch the physical situation to guide your problem-solving. This time the ampèrian loop has a radius less than the radius of the capacitor.	
Find *algebraically* the displacement current through the ampèrian loop located at the radius where you would like to find the magnetic field. The electric field is everywhere parallel to the normal of the surface vector. The electric field is also uniform. Because the area is constant, we need to worry only about changes in the electric field causing a change in the electric flux.	

Determine *algebraically* the electric field that gives rise to the flux through the surface. The electric field produced by the free charge is reduced by the dielectric.	
Substitute this electric field into the expression for the displacement current and solve, still *algebraically*.	
Solve for the actual physical current that flows through the surface. This current is due to the polarization of the bound charges in the mica. As discussed on page 827 in Chapter 24 of the text, the bound charge is related to the free charge.	
Apply Ampère's law *algebraically* with the Maxwell correction to find the magnetic field along surface S. Because of the symmetry in the problem, the magnetic field is everywhere parallel to $d\vec{\ell}$ and is also uniform.	
Relate the polarization and displacement currents to the free charge.	

Substitute values and units to solve for the magnetic field.	
	$B = 3.00 \times 10^{-7}$ T

Check: Because the area is smaller than in Try It Yourself #1, there is less flux through the area, and less physical current. As a result we expect a smaller magnetic field, which we found.

Taking It Further: Assuming a vacuum-filled capacitor, would the magnetic field at a radius less than the capacitor radius still be smaller than the magnetic field just at the capacitor radius? Explain.

30.2 Maxwell's Equations

In a Nutshell

By modifying Ampère's law to include time-varying situations, Maxwell fitted the final piece of the electromagnetic puzzle into place. Gauss's law is a mathematical statement that electric field lines may begin and end only at electric charges; Gauss's law for magnetism is a mathematical statement that magnetic field lines never begin or end, and so magnetic monopoles do not exist. Faraday's law states that a time-varying magnetic field generates a nonconservative electric field, and Ampère's law with the Maxwell correction states that a magnetic field can be generated either by currents or by time-varying electric fields.

Fundamental Equations

Gauss's law

$$\oint_S \vec{E} \cdot d\vec{A} = \oint_S E_n \, dA = \frac{Q_{\text{inside}}}{\epsilon_0}$$

Gauss's law for magnetism

$$\oint_S \vec{B} \cdot d\vec{A} = \oint_S B_n \, dA = 0$$

Faraday's law

$$\oint_C \vec{E} \cdot d\vec{\ell} = -\frac{d}{dt} \int_S B_n \, dA = -\int_S \frac{\partial B_n}{\partial t} \, dA$$

Ampère's law

$$\oint_C \vec{B} \cdot d\vec{\ell} = \mu_0 (I + I_d), \text{ where } I_d = \epsilon_0 \int_S \frac{\partial E_n}{dt} \, dA$$

30.3 The Wave Equation for Electromagnetic Waves

In a Nutshell

Maxwell's equations predict the existence of waves that consist of oscillating electric and magnetic fields. When traveling through free space, both the magnetic and electric fields are solutions to the **wave equation** of the form $\dfrac{\partial^2 \vec{E}}{\partial x^2} = \dfrac{1}{c^2}\dfrac{\partial^2 \vec{E}}{\partial t^2}$ for a wave traveling in the $+x$ direction.

The electromagnetic wave described by the wave equation for both $\vec{E}$ and $\vec{B}$ is a **plane wave**—that is, both $\vec{E}$ and $\vec{B}$ are perpendicular to the direction of travel. In addition, $\vec{E}$ is perpendicular to $\vec{B}$. The magnitudes of the electric and magnetic fields for an electromagnetic wave are related by $E = cB$, and the speed of the wave in vacuum is given by $c = 1/\sqrt{\mu_0 \epsilon_0} = 3.00 \times 10^8$ m/s.

Because the electric and magnetic fields are perpendicular to the direction of propagation, electromagnetic waves are transverse and can be polarized. Electromagnetic waves traveling parallel to the x axis are said to be **linearly polarized** if, at any fixed point, the tip of the $\vec{E}$ vector moves back and forth in a line perpendicular to the x axis; they are **circularly polarized** if, at a fixed point, the tip of the $\vec{E}$ vector moves in a circle whose plane is perpendicular to the x axis.

Physical Quantities and Their Units

Speed of electromagnetic waves in vacuum $\qquad\qquad c = \dfrac{1}{\sqrt{\mu_0 \epsilon_0}} = 3.00 \times 10^8$ m/s

Important Derived Results

Wave equation for $\vec{E}$ $\qquad\qquad\qquad\qquad\qquad \dfrac{\partial^2 \vec{E}}{\partial x^2} = \dfrac{1}{c^2}\dfrac{\partial^2 \vec{E}}{\partial t^2}$

Wave equation for $\vec{B}$ $\qquad\qquad\qquad\qquad\qquad \dfrac{\partial^2 \vec{B}}{\partial x^2} = \dfrac{1}{c^2}\dfrac{\partial^2 \vec{B}}{\partial t^2}$

Field magnitude relationship for electromagnetic waves $\qquad E = cB$

Common Pitfalls

> ➤ The magnitudes of the electric and magnetic fields are related by $E = cB$ for an electromagnetic wave only. This relationship does not hold for static fields.

3. TRUE or FALSE: The $\vec{E}$ and $\vec{B}$ fields of electromagnetic waves in free space (regions where no matter is present) are in phase.

4. It is easier to verify that a time-varying magnetic field produces an electric field (electromagnetic induction) than to verify that a time-varying electric field produces a magnetic field. Why?

Try It Yourself #3

Show by direct substitution that the wave function $\vec{E} = E_0 \left[\cos \left(kx - \omega t \right) + \sin \left(kx - \omega t \right) \right] \hat{k}$ satisfies the wave equation, given that $\omega = kc$.

Picture: Substitute the wave function into the wave equation and demonstrate that you end up with the same expression on both sides of the wave equation.

30.4   Electromagnetic Radiation

In a Nutshell

Electromagnetic waves exist over a broad range of frequencies, from radio waves with frequencies of 10^6 Hz or less to gamma rays with frequencies of more than 10^{20} Hz. Visible light is a narrow band of frequencies in this broad spectrum, in the range from 4.3×10^{14} Hz to 7.9×10^{14} Hz. Before the development of Maxwell's equations, the connection between optics (the study of light) and electricity and magnetism was unknown. Maxwell's equations resulted in the unification of these two previously distinct branches of physics.

A common broadcast antenna is an **electric dipole antenna**, which consists of two conducting rods, one pointing upward and one pointing downward as shown. The charges on the upper and lower rods are always opposite and are periodically exchanged. Such an antenna produces a wave that is linearly polarized with the polarization parallel with the antenna. (The polarization direction of an electromagnetic wave is defined to be in the direction of the electric field vector.) The angular distribution of the intensity of the radiation broadcast by an electric dipole antenna is given by the relation $I = I_0 \sin^2 \theta$, where θ is the angle between the propagation direction and the antenna rods. In this figure the angular dependence of the intensity distribution is illustrated by the oval curve, which is a plot of the relative intensity on the surface of a large sphere centered on the antenna. The length of the arrow is proportional to the intensity broadcast in the direction indicated by the arrow.

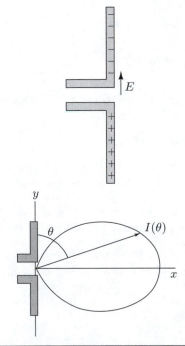

An identical dipole antenna can be used to detect the radiation. Alternatively, a loop antenna, which is commonly used to detect UHF television waves, detects radiation in accordance with Faraday's law. This means that changes in the magnetic flux through the loop induce a current in it.

The **electromagnetic energy density** u of an electromagnetic wave is the sum of the electric energy density u_e and the magnetic energy density u_m.

The **intensity of a wave**—the power delivered per unit area—is the product of the energy density and the velocity. An associated vector quantity is the Poynting vector $\vec{S} = \vec{E} \times \vec{B}/\mu_0$. The Poynting vector's magnitude is the instantaneous intensity of the wave, and its direction is the direction of the wave's propagation.

The magnitude of the momentum p carried by an electromagnetic wave is $1/c$ times the energy U carried by the wave. Consider an electromagnetic wave incident on some surface. If the surface is nonreflective, the wave's energy and momentum are both transferred to the surface. Momentum transferred per unit time is force. Thus, the intensity divided by c is a force per unit area, which is a pressure. This pressure is called radiation pressure P_r. If the wave is reflected, then the radiation pressure equals $1/c$ times the sum of the incident intensity and the reflected intensity.

Important Derived Results

Energy density of electromagnetic waves $\qquad u = u_e + u_m = \epsilon_0 E^2 = \dfrac{B^2}{\mu_0} = \dfrac{EB}{\mu_0 c}$

Poynting vector $\qquad \vec{S} = \dfrac{\vec{E} \times \vec{B}}{\mu_0}$

Intensity of electromagnetic waves $\qquad I = \dfrac{E_{rms} B_{rms}}{\mu_0} = \dfrac{E_0 B_0}{2\mu_0} = \left| \vec{S} \right|_{av}$

Momentum carried by an electromagnetic wave $\qquad p = \dfrac{U}{c}$

Radiation pressure $\qquad P_r = \dfrac{I}{c} = \dfrac{E_0 B_0}{2\mu_0 c} = \dfrac{\epsilon_0 E_0^2}{2\mu_0 c^2} = \dfrac{B_0^2}{2\mu_0}$

Electric dipole antenna angular-intensity distribution $\qquad I = I_0 \sin^2 \theta$

Common Pitfalls

> ➢ Don't forget: Radiation pressure is exerted on a surface by a beam leaving the surface as well as by one incident on the surface.

> ➢ The magnitude of the Poynting vector is the instantaneous intensity, not the average intensity.

5. TRUE or FALSE: The total energy density in an electromagnetic wave in free space is equally divided between that of the electric field and that of the magnetic field.

6. The radiation pressure exerted on a perfectly reflecting surface is $2I/c$. This is twice the value given by the equation for radiation pressure. Why is there a factor of 2 in this relation?

Try It Yourself #4

A local radio station broadcasts with 10.0 kW of power at a frequency of 1340 kHz. If you are detecting this signal with a 30.0-cm straight-wire antenna a distance of 5.00 km from the antenna, what is the maximum voltage that goes to the amplifier and tuner? Assume that the antenna radiates uniformly in all directions (it is not a dipole antenna). How must your antenna be oriented relative to the incoming signal?

Picture: The electric field strength can be determined from the intensity of the signal. Assume the electric field strength is uniform along the length of the wire.

Solve:

Determine the intensity of the signal at your location. The antenna must radiate into a full sphere, 5 km away.	
Find the electric field strength at the location of the receiving antenna by relating the field to the intensity.	
For a uniform electric field, $\Delta V = EL$. The maximum voltage will occur for the maximum value of the electric field, and when the electric field vector is parallel to the wire.	$V_{\max} = 46.5$ mV

Check: All the units are correct, and the values are reasonable.

Taking It Further: How can you increase the magnitude of the voltage generated by the antenna?

Try It Yourself #5

For the broadcast antenna in the previous example, what is the smallest-radius circular loop antenna you would need to produce the same size signal? How must this antenna be oriented relative to the field to achieve maximum sensitivity?

Picture: A loop antenna is sensitive to changes in magnetic flux, via Faraday's law, rather than the electric field.

Solve:

Determine the intensity of the signal at your location. The antenna must radiate into a full sphere, 5 km away.	
Find the magnetic field strength at the location of the receiving antenna by relating it to the intensity.	
Find an *algebraic* expression for the changing magnetic flux through a circular antenna of area A. Remember that the magnetic field of an electromagnetic wave varies sinusoidally.	
The maximum voltage signal will occur when the rate of change of the flux is at a maximum. Find an *algebraic* expression for the maximum value of the rate of change of flux.	

The maximum rate of change of flux is supposed to produce a signal of 46.5 mV. Use this to solve for the radius of the circular loop.	
	$r = 1.85$ m

Check: This is a pretty large radius, but not unseemly, and the units are correct.

Taking It Further: How can the signal generated by a circular loop antenna be increased?

QUIZ

1. TRUE or FALSE: The displacement current is the rate of flow of free charges.

2. TRUE or FALSE: Maxwell's equations apply only to $\vec{E}$ and $\vec{B}$ fields that vary sinusoidally with time.

3. Which of Maxwell's equations denies the existence of magnetic monopoles?

4. Many television sets come with a loop antenna for the UHF channels. How should that antenna be oriented relative to the direction of the transmitting antenna, assuming that the wave is transmitted by an electric dipole antenna?

5. If all you know is the direction of the electric and magnetic fields that create an electromagnetic wave, can the propagation direction of the wave be determined? How?

6. At the top of Earth's atmosphere the intensity of sunlight is 1.35 kW/m^2. Find the area of a 100 percent reflecting surface upon which the force of sunlight is 10^{-5} N (about the weight of a mosquito). Speculate as to whether you can feel the radiation pressure from the sunlight on your skin.

7. A long, straight wire of radius a has a resistance per unit length b and is carrying a steady current I. Calculate the Poynting vector at the surface of the wire, and from it calculate the energy flow per unit time through the surface into a length L of the wire. How does this compare with the power being dissipated in the wire?

Chapter 31

Properties of Light

31.1   The Speed of Light

In a Nutshell

The **speed of light** can be deduced by measuring the time light requires to traverse a known distance. This speed is very high, so unless the distance traveled is large the time interval is very small. The first evidence that the speed of light is finite came from Rmer's seventeenth-century observations of the eclipses of the moons of Jupiter. In the nineteenth century the first laboratory measurements of the speed of light were made by using a toothed wheel, a rotating mirror, or a similar device to periodically interrupt the light reflected from a distant mirror. Current experimental measurements have excellent agreement with Maxwell's theoretical expression for the speed c of electromagnetic waves in empty space, $c = 1/\sqrt{\mu_0 \epsilon_0} = 3.0010^8$ m/s.

Physical Quantities and Their Units

Speed of light in a vacuum
$$c = \frac{1}{\sqrt{\mu_0 \epsilon_0}} = 3.0010^8 \text{ m/s}$$

Try It Yourself #1

In Foucault's experiment to measure the speed of light (Figure 31-3 on page 1057 of the text), suppose that the distance between the fixed mirror and the rotating mirror is $L = 10.0$ km. What is the lowest rate of rotation of the octagonal mirror, in rev/min, at which you could see light in the viewing telescope?

Picture: The time it takes the light to travel from the octagonal mirror to the fixed mirror and back must be the time required for one-eighth of a revolution of the octagonal mirror.

Solve:

Find an *algebraic* expression for the time required for light to travel to the fixed mirror and back.	
Find an *algebraic* expression for the angle through which the mirror rotates during this time. (You may want to review the kinematics of rotational motion. Make sure your expression is in terms of ω.	

Solve for the angular velocity ω. Make sure you convert your units.	
	$\omega = 113\,000$ rev/min

Check: The units work out properly.

Taking It Further: Do you think Foucault really used an octagonal mirror? Explain.

Try It Yourself #2

The radius of the orbit of Mars around the Sun is $2.28 10^{11}$ m and that of Earth's orbit is $1.50 10^{11}$ m. Exploratory rovers on Mars use radio waves to communicate with Earth. What is the maximum value of the time delay for their signal to reach the Earth? What is the minimum value of the delay?

Picture: The time delay is simply distance divided by speed.

Solve:

Calculate the maximum distance, which will be the sum of the two radii. This occurs when the planets are on opposite sides of the Sun.	
Calculate the maximum time.	$t = 1260$ s $= 21.0$ min
Calculate the minimum distance.	

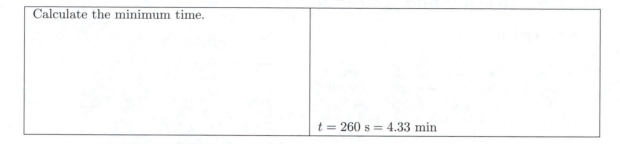

Calculate the minimum time.	
	$t = 260$ s $= 4.33$ min

Check: The units all check out properly.

Taking It Further: Why must exploratory rovers on Mars be at least partially autonomous—that is, not require continuous instructions from Earth?

31.2 The Propagation of Light

In a Nutshell

The propagation of light waves in a given physical situation can be modeled by **Huygens's principle**: "Each point on a wavefront can be considered as a point source of secondary hemispherical wavelets that propagate with the same speed and wavelength as the primary wave. At some later time, the primary wavefront is the envelope of all the secondary wavelets." (An envelope is a surface tangent to each of a family of surfaces.)

An alternative description of the propagation of light is given by **Fermat's principle**: "The path taken by light from one point to another in a given situation is that for which the travel time is a minimum." Fermat's principle is a statement of the conditions necessary for the constructive interference of a propagating wavefront with itself. Fermat's principle can be stated mathematically as $dt/dx = 0$, where t is the time for light to travel along a path and x is a parameter that specifies the path; that is, the travel time t is a minimum if $dt/dx = 0$.

31.3   Reflection and Refraction

In a Nutshell

When electromagnetic waves encounter a boundary between two transparent materials or between a transparent material and an opaque material, **reflection** occurs. Reflection from a smooth surface is **specular** (mirror-like), whereas reflection from a rough surface is **diffuse**. In the case of specular reflection, the directions of propagation of the incident and the reflected wavefronts make equal angles, the **angle of incidence** θ_1 and **angle of reflection** θ_1', with respect to the normal to the surface of the reflecting boundary: $\theta_1 = \theta_1'$.

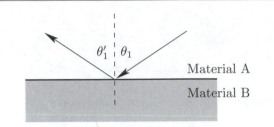

The major optical characteristics of a transparent material are summarized by its index of refraction n, which is the ratio of the speed of light in a vacuum to that in the material: $n = c/v$, where v is the speed of light in the material. The speed of light in most materials is less than that in a vacuum, so indices of refraction are generally greater than one.

The fraction of the light intensity that is reflected at a boundary depends on the indices of refraction in the two media, the angle at which light strikes the boundary, and the state of polarization of the incident light. In the special case of normal incidence of light with an intensity of I_0 the reflected intensity is $I = I_0(n_1 - n_2)^2/(n_1 + n_2)^2$.

When electromagnetic waves are incident on a boundary between transparent media, some of the energy is transmitted across the boundary. The change in the direction of propagation of the waves as they cross the boundary is called **refraction**. Refraction at a boundary depends on the wave speeds (or alternatively, the refractive indices) of both media. The relation between the directions of propagation of the incident and transmitted waves, called the **law of refraction** (or **Snell's law**), is $n_1 \sin\theta_1 = n_2 \sin\theta_2$, where n_1 and n_2 are the indices of refraction and θ_1 and θ_2 are the angles between the directions of propagation and the normal to the boundary. The angle between the direction of propagation of the refracted (transmitted) wave and the normal is called the **angle of refraction**.

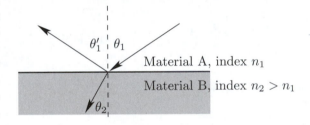

When light propagating in a medium is incident on a boundary between that medium and a medium of lower refractive index, for sufficiently large angles of incidence the law of refraction predicts the impossible—that the sine of the angle of refraction is greater than 1. For these large angles of incidence no refraction occurs, and the light is totally reflected back into the first medium. This **total internal reflection** explains why light traveling along optical fibers remains in the fibers even when they bend, and can also explain mirages. The angle of incidence θ_1 for which the law of refraction predicts that the sine of the angle of refraction equals 1 (that the angle of refraction equals 90°) is called the **critical angle**: $\sin\theta_c = n_2/n_1$. Total internal reflection occurs for all angles greater than the critical angle.

The speed at which light propagates in a material varies with the frequency of the light. As a result, when white light is refracted at a boundary, each frequency (color) is refracted through a different angle. Thus the light is separated into its component colors. The name given to this effect is **dispersion**. The dispersion of white light in water droplets in the atmosphere gives rise to the rainbow.

Important Derived Results

Index of refraction
$$n = \frac{c}{v}$$

Law of reflection
$$\theta_1 = \theta_1'$$

Law of refraction (Snell's law)
$$n_1 \sin \theta_1 = n_2 \sin \theta_2$$

Critical angle for total internal reflection
$$\sin \theta_c = \frac{n_2}{n_1}$$

Intensity of reflected light at normal incidence
$$I = \left(\frac{n_1 - n_2}{n_1 + n_2} \right)^2 I_0$$

Common Pitfalls

> The index of refraction is the speed of light in a vacuum divided by its speed in a material; you are likely to invert this division if you aren't careful! Remember that the index of refraction typically is a dimensionless number greater than 1.
> Light passes from one medium into another with its frequency unchanged. When the speed of propagation changes, the wavelength changes in such a way that the ratio of speed to wavelength remains the same.
> Total internal reflection occurs only for light propagation *from* the medium with the lower speed of light into that with the higher speed—that is, from the medium of higher refractive index into that with the lower refractive index.
> The laws of reflection and refraction always use angles measured with respect to the normal of the surface, *not* angles measured from the surface.

1. TRUE or FALSE: If light is incident normally on an air–glass boundary, the intensity of the reflected light is greater than the intensity of the transmitted light.

2. In some situations light propagates through a medium with a continuously varying index of refraction. The refractive index of a gas, for example, is proportional to its density, so the index of refraction of the atmosphere decreases with increasing altitude. What would the path of a light ray from the Sun look like in such a situation? Explain.

Try It Yourself #3

Light traveling in air is incident normally on a slab of flint glass of refractive index 1.63. Assuming no absorption, what fraction of the incident light energy passes through the slab?

Picture: How much light energy is reflected? The rest must be transmitted.

Solve:

Calculate the intensity of the light reflected at the air–glass interface when the light enters the glass.	
Determine the intensity of light transmitted through the air–glass interface.	
Calculate the intensity of the light reflected at the glass–air interface as the light leaves the glass.	
Determine the intensity of the transmitted light through the entire slab.	$I_{\text{transmitted}} = 0.889 I_0$

Check: The total transmitted intensity is less than the incident intensity, as expected. In this analysis we did not consider what happened to the light reflected at the glass–air interface. To do this problem with complete rigor, however, we need to consider that the light reflected from that interface gets reflected many times within the glass slab, which adds a small amount to the output intensity.

Taking It Further: What is the intensity of the reflected light? Explain.

Try It Yourself #4

The object in the figure is a distance $d = 0.850$ m below the surface of clear water. How far from the end of the dock, distance D in the figure, must the object be if it cannot be seen from any point above the surface? The index of refraction of water is 1.33.

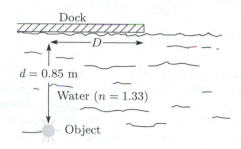

Picture: If the object cannot be seen from a point above the surface, then no light from the object can be transmitted through the surface into the air. Therefore, total internal reflection must be occurring at the water-air interface. Assume the surface of the water is perfectly flat (no waves) to make the calculations easier.

Solve:

On the figure, show the distances and angles involved. The most important ray is one from the object to the end of the dock.	
Use the law of refraction to solve for the critical angle, which is the smallest angle for which total internal reflection can occur.	
Light from the object to the end of the dock must be incident at the critical angle or larger to not be seen. Use this angle to solve for D.	$D = 0.969$ m

Check: The units check out, and this distance seems reasonable.

Taking It Further: If you could change the index of refraction of the water, how would you change it so that the object *could* be seen at any distance from the dock? Explain.

31.4   Polarization

In a Nutshell

Light is a transverse wave and thus can be **polarized**. A wave is polarized if the plane in which its electric field oscillates is fixed or rotates in a simple way; it is unpolarized if the electric field direction varies randomly. An electromagnetic wave produced by a source that is small compared with the wavelength is usually polarized. The polarization state of light is affected by its interactions with matter in four distinct ways: absorption, reflection, scattering, and passing through a birefringent (doubly refracting) material.

A sheet of Polaroid absorbs any component of the electric field that is perpendicular to the transmission axis of the sheet. Thus any light transmitted through such a sheet is polarized in a plane parallel with the sheet's transmission axis. If unpolarized light is incident on an ideal polarizing sheet, the intensity of the transmitted light is half the intensity of the incident light. If linearly polarized light is incident on an ideal polarizing sheet, the intensity of the transmitted light I is related to the intensity of the incident light I_0 according to Malus's law: $I = I_0 \cos^2 \theta$, where θ is the angle between the transmission axis and the direction of polarization of the incident light. Polarizing sheets are frequently used in pairs. The first sheet, called a **polarizer**, polarizes light; the second sheet, called an **analyzer**, determines the plane of polarization of polarized light.

When light shines on a small object such as a molecule or a dust particle, the electric field of the light causes the electrons in the object to oscillate. These oscillating charges, which constitute an electric dipole antenna, emit light, as described in Chapter 30 on pages 1042–1044 of the text. The light emitted by an electric dipole antenna is linearly polarized.

When unpolarized light is reflected from the boundary between two transparent media such as air and water, the reflected light is partially polarized. When the angle between the reflected and refracted ray is 90°, the reflected light is completely polarized parallel to the plane of the surface, and the angle of incidence is called the **polarizing angle** θ_p. This condition, along with the laws of reflection and refraction, give us the relation (Brewster's law) between the polarizing angle and the indices of refraction: $\tan \theta_p = n_2/n_1$, where the incident wave is in the medium with refractive index n_1.

Important Derived Results

Malus's $I = I_0 \cos^2 \theta$

Polarizing (Brewster's) angle $\tan \theta_p = \dfrac{n_2}{n_1}$

Common Pitfalls

> Light is polarized (or more precisely, its state of polarization is affected) by reflection at any angle between 0 and 90°. The polarizing angle (Brewster's angle) is the angle at which all the reflected light is linearly polarized, and its direction of polarization is parallel to the reflecting surface.

3. TRUE or FALSE: If light falls on a boundary at the polarizing angle, both the refracted and the reflected light are completely polarized.

4. Which way should the transmission axis be oriented in antiglare polarizing sunglasses? Explain.

Try It Yourself #5

The polarizing angle for light that passes from water, $n = 1.33$, into a certain plastic is $61.4°$. What is the critical angle for total internal reflection of the light passing from this plastic into air?

Picture: Use the polarizing angle to find the index of refraction of the plastic. Use that result to find the critical angle.

Solve:

Draw a sketch showing light incident on the plastic at the polarizing angle.	
Use the polarizing angle to find an *algebraic* expression for the index of refraction of the plastic.	
Draw a sketch showing the total internal reflection.	
Using the expression for the index of refraction found in the second step, determine the critical angle for light traveling from plastic to air.	$\theta_c = 24.2°$

Check: This angle seems reasonable.

Taking It Further: What is the polarizing angle for light traveling from the plastic into air? Is the polarizing angle always less than the critical angle, if there is one? Explain.

Try It Yourself #6

Unpolarized light of intensity I_0 is incident on two ideal polarizing sheets that are placed with their transmission axes perpendicular to each other. An additional polarizing sheet is then placed between these two, with its transmission axis oriented at 30° to that of the first. What is the intensity of the light passing through the stack of polarizing sheets?

Picture: The first polarizing sheet will reduce the total intensity in half because it will block out half the unpolarized light. After passing through the first sheet, the light is linearly polarized, so Malus's law can be used to calculate the fraction of the incident intensity transmitted by the second and third sheets, successively.

Solve:

Determine the amount of light transmitted by the first polarizer that is incident on the second polarizer.	
Determine the amount of light that gets through the second polarizing sheet to the third sheet, using Malus's law. Because of the orientation of the second sheet, the light leaving the second sheet will have a polarization rotated 30° with respect to the incoming polarization.	
Use Malus's law again to find the amount of light that passes through the third polarizing sheet, which is now at 60° with respect to the polarized light incident upon it.	$I = \dfrac{3I_0}{32}$

Check: The transmitted intensity is less than the incident intensity. This problem illustrates a remarkable property of polarization: by adding a third element to a crossed polarizer and analyzer, you can actually get more light transmitted through the system than without the third element!

Taking It Further: What orientation of the middle sheet enables the three-sheet combination to transmit the greatest amount of light? Explain.

31.5   Derivation of the Laws of Reflection and Refraction

In a Nutshell

The laws of reflection and refraction can be derived from either Fermat's principle or by using Huygens's construction.

31.6   Wave–Particle Duality

In a Nutshell

The wave nature of light was first demonstrated persuasively by Thomas Young and his contemporaries. Young observed the interference pattern of two coherent light sources produced by illuminating a pair of narrow, parallel slits with a single source. The wave theory of light culminated in 1860 with Maxwell's prediction of electromagnetic waves.

The particle nature of light was first proposed by Albert Einstein in 1905 in his explanation of the photoelectric effect. A particle of light called a photon has energy E that is related to the frequency f and wavelength λ of the light wave by the Einstein equation $E = hf = hc/\lambda$, where $h = 6.626 10^{-34}$ J $\cdot$ s is known as Planck's constant.

The propagation of light can be described both by its wave properties, via classical electrodynamics, and by its particle properties, via quantum electrodynamics. However, energy and momentum exchanges between light and matter are always both localized and quantized. This duality between light's particle nature and wave nature is a general property of nature. For example, electrons and other so-called particles can exhibit wave properties under certain conditions.

Physical Quantities and Their Units

Planck's constant $\qquad\qquad\qquad\qquad\qquad\qquad h = 6.626 10^{-34}$ J $\cdot$ s $= 4.136 10^{-15}$ eV $\cdot$ s

Fundamental Equations

Einstein's equation for photon energy $\qquad\qquad E = hf = \dfrac{hc}{\lambda}$

Common Pitfalls

> The laws of reflection and refraction can be explained by *either* a wave *or* a particle theory of light. It is the implications of one model or the other that enables us to experimentally distinguish between the two models. For example, when light travels from air to water the light bends toward the normal. Newton's particle theory explains this by asserting light consists of particles that travel faster in water than in air. Huygens's wave theory explains it by asserting light consists of waves that travel slower in water than in air. By measuring the speed of light in both air and water, Foucault demonstrated that light travels more slowly in water than in air. This observation discredited Newton's particle theory of light.

> The wave–particle debate about the nature of light was resolved in the early part of the 20th century. In the current model, light has both wave and particle attributes. Light exhibits primarily wave characteristics in propagation and particle characteristics in interactions with atoms, molecules, and nuclei, such as absorption and emission.

31.7   Light Spectra

In a Nutshell

Using a prism, Newton demonstrated that the spectrum of sunlight was composed of all colors, with the intensity of the different colors being approximately equal. Such a combination of colors produces the sensation of white light. Because the spectrum of sunlight contains a continuous range of wavelengths, it is called a **continuous spectrum**. The light emitted by the atoms in low-pressure gases contains only a discrete set of wavelengths which, upon refraction (or diffraction through an array of parallel slits), produce a **line spectrum**.

31.8   Sources of Light*

In a Nutshell

Many common sources of visible light are transitions of the outer electrons in atoms. Normally an atom is in its ground state with its electrons at their lowest allowed energy levels. The lowest energy electrons are closest to the nucleus and are tightly bound, forming a stable inner core. The one or two electrons in the highest energy states are much farther from the nucleus and are relatively easily excited to vacant higher energy states. These outer electrons are responsible for the energy changes in the atom that result in the emission or absorption of visible light.

When an atom collides with another atom or with a free electron, or absorbs electromagnetic energy, the outer electrons can be excited to higher energy states. After about 10^{-8} s these outer electrons spontaneously make transitions to lower energy states with the emission of a photon. This process, called **spontaneous emission**, is random; the photons emitted from two different atoms are not correlated, even though their frequencies may be the same. The frequency of the light wave is related to the energy by the Einstein equation, $\Delta E = hf$. The wavelength of the emitted light is then $\lambda = hc/|\Delta E|$, where $|\Delta E|$ is both the energy difference between the emitting atom's initial and final states and the energy of the emitted photon. Because electrons reside only in specific energy levels in atoms, they can make only a particular set of transitions from high to low energy states. Thus the **emission spectrum** of light from single atoms or atoms in low-pressure gases consists of a set of sharp discrete lines—that is, specific wavelengths or photon energies—that are characteristic of the element.

When atoms are close together and interact strongly, as in liquids and solids, the energy levels of the individual atoms are spread out into energy bands, resulting in an essentially continuous range of energy band levels. Thus when the outer electrons of these atoms are excited, there is a **continuous spectrum** of possible transition energies, which results in a continuous emission spectrum instead of a discrete line spectrum.

In incandescent objects such as the filament of a light bulb or the surface of the Sun, the electrons are randomly excited by frequent atom–atom collisions. This results in a broad spectrum of thermal radiation, and the light emitted by such incandescent objects is a continuous spectrum. The thermal radiation emitted by an object below about 1000 K is concentrated in the infrared region and is therefore not visible. The surface of the Sun, which has a temperature of 6000 K, emits a continuous spectrum. Over the visible range, the intensity of the Sun's radiation is nearly uniformly distributed.

Radiation is emitted when an atom makes a transition from an excited state to a state of lower energy; radiation is absorbed when an atom makes a transition from a lower state to a higher state. When atoms are irradiated with a continuous spectrum of radiation, the transmitted spectrum shows dark lines corresponding to absorption of light at discrete wavelengths. The **absorption spectra** of atoms were the first line spectra observed. Since atoms and molecules at temperatures typical to our environment are in either their ground states or low-lying excited states, absorption spectra usually have fewer lines over a given range of wavelengths than do emission spectra.

---

*Optional material

If the energy of the incoming photon is too small to excite the atom, the atom remains in its ground state and the photon is said to be scattered. Since the incoming and outgoing, or scattered, photons have the same energy, the scattering is said to be elastic. If the wavelength of the incident light is large compared with the size of the atom, the scattering can be described in terms of classical electromagnetic theory and is called **Rayleigh scattering** after Lord Rayleigh, who worked out the theory in 1871. The probability of Rayleigh scattering varies as $1/\lambda^4$. This means that when sunlight is scattered by air molecules, the blue light is scattered much more readily than red light, which accounts for the bluish color of the sky. When you look directly at the sun during sunrise or sunset, the blue light has been removed by Rayleigh scattering. Only the longer wavelengths remain to enter your eye, which accounts for the reddish-orange color that you see.

If the energy of the incident photon is just equal to the difference in energy between the ground state and the first excited state of the atom, the atom makes a transition to its first excited state. After a short delay, it decays by spontaneous emission back to the ground state, and a photon is emitted whose energy is equal to that of the incident photon. This multistep process is called **resonance absorption**.

Inelastic scattering occurs when the incident photon has more than enough energy to cause the atom to make a transition to an excited state. A photon with energy hf' is scattered, but its energy is less than that of the incident photon, hf, by ΔE, the energy absorbed by the atom as it transitions to its excited state. Inelastic scattering of light from molecules was first observed by the Indian physicist C. V. Raman and is often referred to as **Raman scattering**.

If the energy of the incident photon is great enough to excite the atom past its first excited state to one of its higher excited states, the atom then loses its energy by spontaneous emission as it makes one or more transitions to lower energy states. A common example occurs in **fluorescence**, where the atom is excited by ultraviolet light and emits visible light as it returns to its ground state. Since the lifetime of a typical excited atomic energy state is of the order of 10^{-8} s, fluorescence is often a short-lived process. However, some excited states have much longer lifetimes, of the order of milliseconds or occasionally seconds or even minutes. Such a state is called a **metastable state**. **Phosphorescent materials** have very long-lived metastable states, and so emit light long after the original excitation.

In the above cases of spontaneous emission, the phase of the emitted photon is not correlated with the phase of the incident photon. However, **stimulated emission** results in photons traveling in phase with one another. In stimulated emission an atom or molecule is initially in an excited state of energy E_2 and the energy of the incident photon is equal to $E_2 - E_1$, where E_1 is the energy of a lower state or the ground state. The oscillating electromagnetic field associated with the incident photon stimulates the excited atom or molecule, which then emits a photon in the same direction as the incident photon and in phase with it. In stimulated emission, the phase of the light emitted from one atom is related to that emitted by every other atom, so the resulting light is **coherent**.

The **laser** (**l**ight **a**mplification by **s**timulated **e**mission of **r**adiation) is a device that produces a strong beam of coherent photons by stimulated emission. For a laser to work, we must first have a population inversion—that is, more of the atoms have to be in an excited, but metastable, state of energy E_2 than in a lower state E_1 so that stimulated emission will dominate over absorption. (Normally, the atoms in a system at temperatures typical to our environment will be in the ground state.) A population inversion can be achieved by a variety of methods in which atoms are "pumped" up to energy levels of higher energy than E_2 by absorbing energy supplied by an auxiliary source. The atoms then decay down to the metastable state E_2 by either spontaneous emission or nonradiative transitions such as those due to collisions. When some of the atoms at E_2 decay to the lower state by spontaneous emission, they emit photons. These photons stimulate other atoms in the same metastable state to emit photons with the same energy in the same phase and direction. The emitted photons and incident photons combine to form a more intense light beam.

If a laser medium is placed between two mirrors, each successive bounce of the light stimulates additional coherent emission of photons, which increases the intensity of the light. This system is called a **laser cavity**. If one of the mirrors is not 100 percent reflecting, so that a small fraction of the light can be transmitted through it, a laser beam is produced.

QUIZ

1. TRUE or FALSE: Light from the sky is polarized because of the polarization associated with the scattering of sunlight by the air molecules.

2. TRUE or FALSE: A ray is a directed line that is perpendicular to the wavefronts of a wave.

3. When light propagates across a boundary into a second transparent medium with a higher index of refraction, its wavelength decreases but its frequency remains the same. Why?

4. Why is the highway so much harder to see when you are driving on a rainy night than on a dry one?

5. Why does the oar you are rowing with appear to be bent at the water surface?

6. A flat glass surface with $n = 1.54$ has a layer of water, $n = 1.33$, of uniform thickness directly above the glass. At what minimum angle of incidence must light in the glass strike the glass–water interface for the light to be totally internally reflected at the water–air interface?

7. A ray of light is incident at an angle of 45° on a slab of glass 1.00 cm thick as shown. The index of refraction of the glass is 1.51. (a) Show that the emergent ray is parallel to the incident ray. (b) Find the lateral displacement d between the incident and emergent rays.

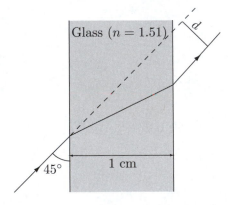

Chapter 32

Optical Images

32.1 Mirrors

In a Nutshell

Light propagates along straight lines through a uniform medium as long as the sizes of the obstacles or apertures it encounters are large compared with the wavelength(s) of the light. This is the domain of geometric (ray) optics. The direction of the light may change at the interfaces between different transparent media.

The statement that "although you see with light, you cannot see light" often startles people. Light cannot be seen in the sense that if you look across the table at your dining companion, you cannot see the light from the candle that passes upward through the space between you and illuminates the ceiling. Anything that you do see distinctly is either an object, a virtual image, or a real image.

A source of light is called an **object** if light radiates or is scattered from each point on its surface. Some objects, like the filament of an incandescent light bulb, generate their own light, whereas others, like the face of your dining companion, merely reflect light diffusely. Things that reflect light specularly, like mirrors, cannot be seen and are not, in this sense of the word, objects.

You see an object because the light from a point on the object that enters your eye is refracted by your cornea and lens to a point on your retina. If light from a point source is reflected from a smooth surface, such as a plane mirror, the light rays enter the eye as shown in the figure just as if they came from a point behind the mirror. This apparent source is a **virtual image** of the actual source—virtual in the sense that light does not actually emanate from the image. The image formed by a plane mirror is at the same distance behind the plane of the mirror as the source of light is in front of it. This image can be viewed by an eye located anywhere in the shaded region of the figure. The image formed by one reflecting surface can be imaged again by another reflection surface.

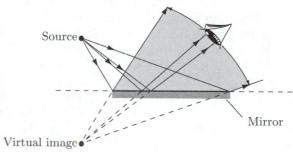

Light from a point source that is reflected by a concave spherical mirror as shown can form a **real image**—real in the sense that, after reflecting from the mirror, the light rays actually converge to the image point and then diverge from it. Viewed by the eye in the figure, the rays diverge from the image point just as if the image point were a point source. All rays emanating from a point source that remain close to (almost parallel to) the symmetry axis of a spherical mirror, or **paraxial rays**, will, after reflection, either converge to or diverge from a point image. The nonparaxial rays from the point source that strike the mirror will, after reflection, cross near but not through the image point, thereby producing a blurred image. How close a ray is to being paraxial determines how close to the image point it passes.

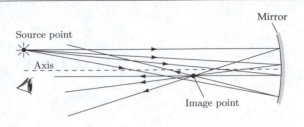

Imaging systems are reversible; that is, if the light from an object located at position A forms an image of the object at location B, then the light from an object located at position B forms an image at location A.

The parallel paraxial rays shown are reflected by a concave spherical mirror to an image point on the **focal plane**, a plane located halfway between the mirror and its center of curvature. If the parallel incident rays are also parallel with the axis, the location of the **image point**, called the **focal point** of the mirror, is on the axis. The **focal length** of the mirror is defined as the distance from the focal point to the center of the mirror, the **vertex**. Thus, the focal length equals one-half the radius of curvature of the mirror. Any point source located closer to the mirror than the focal point forms a virtual image located behind the mirror; that is, paraxial rays from point sources that are located closer to the mirror than the focal length will, on reflection, diverge as if they came from a point behind the mirror.

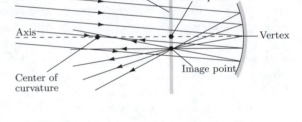

The formation of an image point by a convex mirror is illustrated here.

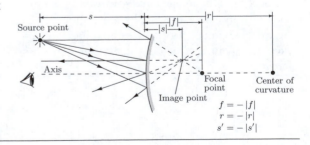

$$f = -|f|$$
$$r = -|r|$$
$$s' = -|s'|$$

We can calculate the location of an image point by using a formula relating it to the location of the source point and the focal length, where s, s', f, and r are the directed distances from the mirror to the source point, image point, focal point, and center of curvature, respectively. The positive direction is toward the source. The sign conventions, that allow the mirror formula to be valid for both concave and convex mirrors are as follows:

1. If the object point is on the same side of the mirror as the incident light, the distance s is positive.
2. If the image point is on the same side of the mirror as the reflected light, the distance from the mirror to the image point is positive; otherwise it is negative.
3. If the center of curvature of the mirror is on the same side of the mirror as the reflected light, the mirror's radius of curvature r and focal length f are positive, otherwise they are negative.

The **mirror equation** is established by applying the law of reflection to paraxial rays.

If all we need is the approximate location of an image, we can locate it quickly by drawing a **ray diagram** of the system and the principal rays. The **principal rays** are easiest to construct. Here we have ray diagrams that show image formation, using the principal rays for a concave and a convex mirror, respectively. The principal rays from the source point are:

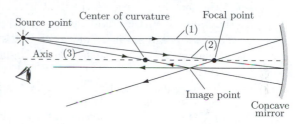

1. The **parallel ray**, incident parallel to the axis of the system, is reflected through the focal point.
2. The **focal ray**, incident on the mirror on a line through the focal point, is reflected parallel to the axis.
3. The **radial ray**, incident on the mirror on a line through the center of curvature, is reflected back on itself.

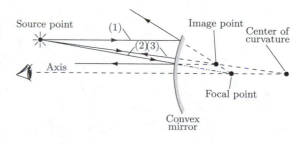

In each figure, an eye has been drawn at a position from which you could view the image. To see an object or image, the light entering your eye must be light that is diverging from points on an object or image.

Each point on an extended object is a point source of light. The image of such an object is the set of image points formed by the reflection of light from the object. For an extended object, the size of the image is usually not the same as the size of the object, and the ratio of the image size to the object size is called the **lateral magnification** of the image, $m = y'/y = -s'/s$, where y and y' are the distances of the object and the image points above the axis, and s and s' are distances from the mirror to the object (the object distance) and from the mirror to the image (the image distance) as shown. This magnification is defined to be positive for an erect image and negative for an inverted image.

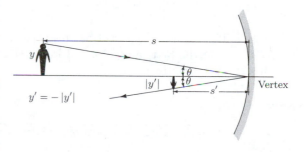

Important Derived Results

Mirror equation
$$\frac{1}{s} + \frac{1}{s'} = \frac{1}{f}, \text{ where } f = \frac{r}{2}$$

Lateral magnification
$$m = \frac{y'}{y} = \frac{-s'}{s}$$

Common Pitfalls

> The formulas for calculating image positions for mirrors must be used with proper signs for all quantities, or the answers will be nonsense. Always keep these sign conventions in mind:
>   1. An object is real, and the object distance is positive, if the object is on the same side of the lens or mirror as the light was traveling *before* it was reflected or refracted; otherwise the object is virtual, and the object distance is negative.
>   2. An image is real, and the image distance is positive, if the image forms on the same side of the lens or mirror as the light is traveling *after* it is reflected or refracted; otherwise the image is virtual, and the image distance is negative.
>   3. The radius of curvature of a reflecting surface is positive if the center of curvature is on the same side of the surface as the light is traveling *after* it is reflected; otherwise it is negative.

> The radius of curvature of a plane surface is infinite.

1. TRUE or FALSE: Paraxial rays are always parallel to the axis of a mirror or lens.

2. A plane mirror seems to invert your image left and right but not up and down. Why is this?

Try It Yourself #1

The opposite walls of a barber shop are covered by plane mirrors, so that multiple images arise from multiple reflections, and you see many reflected images of yourself, receding to infinity. The width of the shop is 6.50 m, and you are standing 2.00 m from the north wall. How far apart are the first two images of you behind the north wall?

Picture: The closest image will be an image of you, a real object. The second closest image will be the image of the first image formed in the south mirror. Remember that plane mirrors have a magnification of exactly 1.

Solve:

Draw a sketch showing the locations of the images.	

Determine the distance behind the north wall the first image of you will appear.	
Determine the distance behind the south mirror the first image of you will appear.	
Determine the distance behind the north mirror the image of the image found in the previous step will be behind the north mirror.	
Determine the difference in the distance between the image of the first step and the image of the previous step.	$d = 9.00$ m

Check: This seems reasonable.

Taking It Further: What is the separation of the first two images of you behind the south wall? Explain.

Try It Yourself #2

An object is 40.0 cm from a concave spherical mirror whose radius of curvature is 32.0 cm. Locate and describe the image formed by the mirror (a) by calculating the image distance and lateral magnification and (b) by drawing a ray diagram. On the ray diagram put an eye in a position from which it can view the image.

Picture: Use the appropriate expressions for the calculations, and use the three principal rays for the ray diagram. Use a ruler to draw the ray diagram as close to scale as possible.

Solve:

Do the ray diagram first. It can be used as a sketch of the situation for the calculations. Place an arrow, to scale, 40 cm from the mirror. Locate the center of curvature and the focal point of the mirror. Then draw the three principal rays.	
Use the mirror equation to find the position of the image.	$s' = 26.7$ cm
Is the image real or virtual?	Since $s' > 0$, the image is real.
Determine the magnification of the image.	$m = -0.668$
Is the image larger or smaller than the object? Upright or inverted?	Since $m < 0$, the object is inverted. Since $|m| < 1$, the image is smaller than the object.

Check: These results seem reasonable, especially given our ray diagram.

Taking It Further: At what location or locations, if any, can an object with a positive s create a virtual image using only a concave lens? Explain.

32.2 Lenses

In a Nutshell

Refraction of light at the boundary between transparent media can also form images. At a spherical boundary, paraxial rays from a point source, upon refraction, either converge to or diverge from a point image. The sign convention for the radius of curvature of a spherical boundary between transparent media is that the radius is positive if the center of curvature is on the same side of the boundary as the refracted light, and negative otherwise. The equation relating the location of the object and the location of the image can be derived using the law of refraction, and is $(n_1/s) + (n_2/s') = (n_2 - n_1)/r$.

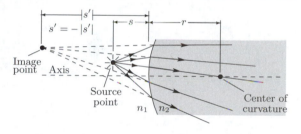

If, after refraction, the rays from a source point continue to diverge as shown, there is a virtual image point on the same side of the boundary as the incident light. For such images the spherical boundary equation gives a negative value for s'. The sign convention for s' is that if the location of the image point is on the same side of the boundary as the refracted light, s' is positive; otherwise it is negative.

The lateral magnification m of a spherical boundary is $m = y'/y = -n_1 s'/(n_2 s)$.

Many applications for imaging by refraction involve **thin lenses**. A lens is a piece of transparent material with spherical surfaces on either side. The focal length of a thin lens depends on the two radii of curvature and the refractive index of the lens material. The **focal point** of a lens is the location at which an incident beam that is parallel with the axis converges to or diverges from. Because light can enter a lens from either side, a lens has two focal points, one on each side. For a thin lens these focal points are equidistant from the lens.

A lens that is thicker at its middle than at its edges and has a higher index of refraction than its surroundings converges light ($f > 0$), and a lens that is thicker at its edges than at its middle and has a higher index of refraction than its surroundings diverges light ($f < 0$). The location of an image point can be calculated using a formula relating it to the location of the source point and the focal length. This relationship is identical to the mirror equation.

The focal length f of a thin lens is related to the index of refraction n of the lens material and to the radii of curvature r_1 and r_2 of the front and back surfaces of the lens by the lens-maker's equation given below.

Images formed by lenses can also be located by drawing ray diagrams; but for lenses the principal rays are

1. the parallel ray,
2. the focal ray, and
3. the undeflected central ray through the center of the lens.

These are illustrated for thin lenses below. Note that although there are two refracting surfaces, for the purposes of ray diagrams of thin lenses, we assume there is only one refracting surface, located in the center of the lens.

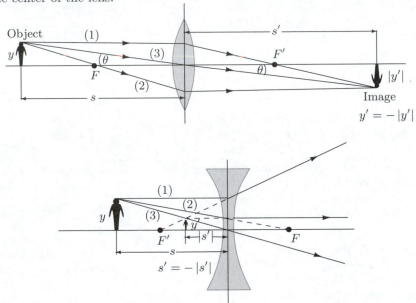

Practical optical instruments often consist of a sequence of several optical elements (thin lenses, refracting surfaces, and mirrors). Such a system is analyzed by taking the image formed by each element as the object that is imaged by the next element. In such an analysis, converging light is sometimes incident on an element. The location of the object (a virtual object) associated with this converging light is on the opposite side of the element from the incident light. The object distance s associated with the location of a virtual object is negative.

The mirror equation discussed above applies to lenses or mirrors, positive or negative lenses, concave or convex mirrors, real or virtual images, provided we follow consistent sign conventions. As a result, the mirror equation is also often referred as the **thin-lens equation** when applied to lenses. The following sign conventions apply to all cases:

1. The object distance s is positive for a real object (an object from which the incident light rays diverge), and is negative for a virtual object (an object toward which the incident light rays converge).
2. The image distance s' is positive and the image is real if the image is on the same side of the element as the reflected or refracted light.
3. The focal length is calculated in terms of the radius of curvature of the reflecting or refracting surface(s). The radius of curvature of a reflecting or refracting surface is positive if the center of curvature is on the same side of the surface as the reflected or refracted light, and negative otherwise. The focal length of a mirror is one-half the radius of curvature, whereas the focal length of a thin lens is related to the radii of curvature and the index of refraction by the lens-maker's formula.

The refracting **power of a lens** is the inverse of its focal length; if the focal length is in meters, the power is given in reciprocal meters, called **diopters** (D).

Physical Quantities and Their Units

Optical power   SI units of diopters (D); $1\ D = 1\ m^{-1}$

Important Derived Results

Image formation at a spherical surface $$\frac{n_1}{s} + \frac{n_2}{s'} = \frac{n_2 - n_1}{r}$$

Lens-maker's equation $$\frac{1}{f} = (n-1)\left(\frac{1}{r_1} - \frac{1}{r_2}\right)$$

Lateral magnification of thin lenses $$m = \frac{y'}{y} = \frac{-s'}{s}$$

Lateral magnification of spherical boundaries $$m = \frac{y'}{y} = \frac{-n_1 s'}{n_2 s}$$

Power of a lens $$P = \frac{1}{f}$$

Common Pitfalls

> The formulas for calculating image positions for lenses must be used with proper signs for all quantities, or the answers will be nonsense. Always keep these sign conventions in mind:
>   1. An object is real, and the object distance is positive, if the object is on the same side of the lens as the light was traveling *before* it was refracted; otherwise the object is virtual, and the object distance is negative.
>   2. An image is real, and the image distance is positive, if the image forms on the same side of the lens as the light is traveling *after* it is refracted; otherwise the image is virtual, and the image distance is negative.
>   3. The radius of curvature of a refracting surface is positive if the center of curvature is on the same side of the surface as the light is traveling *after* it is refracted; otherwise it is negative.
> By convention, we draw ray diagrams with the light proceeding to the right from a source at the left. This is *not* what determines signs. The image distance from a lens is negative if the image is on the side not traversed by refracted light (that is, the same side as the source, and thus is virtual)—not because the image is to the left of the lens.
> A negative lateral magnification corresponds to an inverted image.
> The refracting power of a lens is the inverse of its focal length; a stronger lens has a smaller, not a larger, focal length.

3. TRUE or FALSE: The focal length of a spherical mirror is twice its radius of curvature.

4. Is it possible that a positive lens in air can become a negative lens when immersed in a fluid such as water? If so, in what circumstances?

Try It Yourself #3

A thin lens made of glass of refractive index 1.60 has surfaces with radii of curvature of magnitude 12.0 and 18.0 cm. What are the possible values for its focal length?

Picture: Use the lens-maker's equation, adopting different sign combinations for each of the surfaces to produce different focal lengths.

Solve:

Find the focal length assuming two positive radii of curvature.	
	$f = \pm 60.0$ cm
Find the focal length assuming two negative radii of curvature.	
	$f = \pm 60.0$ cm
Find the focal length if one radius of curvature is positive and the other is negative.	
	$f = \pm 12.0$ cm

Check: These values seem reasonable.

Taking It Further: Sketch a cross-sectional view of the lens for each possible combination, making sure to label the radii of curvature of each surface of the lens and the associated focal length of the entire lens.

Try It Yourself #4

A lens of focal length +15.0 cm is 10.0 cm to the left of a second lens of focal length −15.0 cm. Where is the final image of an object that is 30.0 cm to the left of the positive lens?

Picture: The image produced by the first lens will serve as the object for the second lens. Ignoring the second lens, use the thin-lens equation to find the location of the image produced by the first lens. Using this as the object, find the location of the image produced by the second lens.

Solve:

Draw a sketch of the situation to keep track of the location of the lenses and images.	
Use the thin-lens equation to determine the location of the image produced by the first lens.	
The image created in the previous step serves as a virtual object for the second lens. Use the thin-lens equation a second time to determine the location of the image produced by the second lens. The sign convention is that virtual objects are a negative distance from the lens.	$s' = -60.0$ cm

Check: This answer seems reasonable–it is located to the left of the second lens.

Taking It Further: Is this image real or virtual? How does the image's size and orientation compare to those of the original object? Explain. How must you be oriented to see the image?

32.3   Aberrations*

In a Nutshell

Ideally, all rays from a point source that are incident on an imaging element, such as a mirror, lens, or boundary between transparent media, focus at a single image point. For a spherical imaging element this happens (to a close approximation) for paraxial rays but fails for nonparaxial rays. If all the rays from a point source do not focus at a single point, a blurring of the image, called **aberration**, results. The blurring of the image that results when the object is on the axis is called **spherical aberration**. Additional aberrations, which include **coma** and **astigmatism** result from imaging the light from off-axis point sources. These aberrations result from the law of refraction and the geometry of spherical surfaces. Because of the dependence of the index of refraction on wavelength, images made from light with more than one wavelength are smeared; this type of distortion is called **chromatic aberration**.

32.4   Optical Instruments*

In a Nutshell

The eye is our most familiar and important optical instrument. The pupil is a variable aperture that controls the amount of light admitted to the eye. Light coming into the eye is focused primarily by the front surface (the cornea), and to a lesser extent by the lens, onto the photosensitive real surface (the retina), which is about 2.5 cm behind the cornea. The eye has a focusing power of about 60 diopters (D), with the cornea having 43 D and the lens 17 D. Slight changes in focal length (accommodation) are made by muscles that change the shape of the lens in order to image objects at different distances. In providing accommodation, the focusing power of the lens varies by about 6 D.

The apparent size of an object seen by the eye is determined by the size of the image on the retina; this, in turn, is determined by the angle the object subtends at the eye. For the unaided eye, the greatest apparent (angular) size, and thus the most distinct vision, occurs when the object is at the **near point**, the closest distance from the eye for clear vision. The distance from the eye to the near point varies from eye to eye, but the standard value for this distance is 25 cm. When a converging lens is placed closer to the object than the focal length, a virtual image of the object is formed by the refracted light. Looking into the lens, the viewer sees this image, which has a greater angular size than the object when viewed at the near point of the unaided eye. A lens used in this manner is called a **simple magnifier**. The angular magnification M of a simple magnifier is the ratio of the angular size θ of the virtual image to the angular size θ_0 of the object when viewed by the unaided eye at the near point. When the eye is very close to the lens and when the image is at infinity, $M = \theta/\theta_0 = x_{np}/f$, where f is the focal length of the lens and x_{np} is the distance of the near point to the eye. In microscopes and telescopes, the lens that you look into (the eyepiece or ocular) is a simple magnifier that is used to provide angular magnification of a real image formed by another lens.

There is a limit to the angular magnification that a simple magnifier can provide because a large angular magnification means a short focal length, and, in accordance with the lens-maker's formula, producing a short focal length requires a material with a high index of refraction and surfaces with short radii of curvature. Transparent materials with indices of refraction higher than 2 or so are not available, and surfaces with short radii of curvature require physically small lenses. Unfortunately, small lenses have low light-gathering power and thus form dim images. So we use a compound microscope when a simple magnifier is unable to provide sufficient angular magnification.

---

*Optional material

In a **compound microscope**, a high-power (short focal length) converging lens (the objective) makes a large real image of a small object. This image is the object for a second converging lens (the eyepiece), which is used as a simple magnifier to provide angular magnification. The net angular magnification (magnifying power) of a compound microscope is $M = m_o M_e = -L x_{np}/(f_o f_e)$, where m_o is the lateral magnification associated with the objective lens, M_e is the angular magnification associated with the eyepiece, f_o and f_e are the focal lengths of the objective and eyepiece, respectively, and the tube length L is the distance between the focal points of the objective and the eyepiece.

We use a **telescope** to examine objects that are too distant to be viewed with a simple magnifier. In a telescope, a converging lens with a long focal length forms a real image of the object near its focal point. This image is the object for the eyepiece, which is used as a simple magnifier to provide angular magnification. The net angular magnification (magnifying power) of a telescope is $M = \theta_e/\theta_o = -f_o/f_e$, where θ_o is the angle subtended by the object at the unaided eye and θ_e is the angle subtended by the virtual image (located at infinity) at the eye when looking through the eyepiece. A major purpose of astronomical telescopes is to gather light. Thus an objective lens with a large diameter is desirable.

Important Derived Results

Magnifying power of a simple magnifier

$$M = \frac{\theta}{\theta_e} = \frac{x_{np}}{f}$$

Magnifying power of a compound microscope

$$M = \frac{\theta_o}{\theta_e} = m_o M_e = -\frac{L}{f_o}\frac{x_{np}}{f_e}$$

Magnifying power of a telescope

$$M = \frac{\theta_e}{\theta_o} = m_o M_e = -\frac{f_o}{f_e}$$

Common Pitfalls

➢ The front surface of the eye, the cornea, does most of the focusing. The lens is the apparatus that changes the focus of the cornea–lens system, enabling us to focus on objects over a wide range of distances from the eye.

➢ Angular magnification and lateral magnification can be easily confused. Many optical instruments, such as binoculars, microscopes, and simple magnifiers, form a virtual image that is viewed by the eye. For these instruments, the parameter of interest is the angular magnification (magnifying power). The larger the angular magnification of the instrument, the larger the size of the real image formed on the retina. The size of this real image is proportional not to the lateral size of the virtual image but to its angular size.

➢ Telescope objectives and microscope objectives are easily confused. The lateral magnification of a telescope objective is greater for an objective with longer focal length; the longer the focal length of the objective the larger the size of the real image it produces. The lateral magnification of a microscope objective is greater for an objective with shorter focal length; the shorter the focal length of the objective the larger the size of the real image it produces at the focal plane of the eyepiece.

5. TRUE or FALSE: The image formed by a lens used as a simple magnifier is virtual.

6. Why do some people wear bifocals?

Try It Yourself #5

A farsighted worker needs to read from a word-processor screen that is 50.0 cm from his eyes. His uncorrected near point is 110 cm away. What must the power of the lenses in his reading glasses be to form an image of the screen 15.0 cm beyond the near point? Assume that the reading glasses are at a distance of 2.00 cm from his eyes.

Picture: The glasses must make images of objects very close appear to be further away so the relaxed eye can see them.

Solve:

Draw a sketch of the situation, positioning the eye, the glasses, the object, and the image.	
Use the thin-lens equation to determine the power of the reading glasses. Remember, s and s' are relative to the lens of the glasses, not of the eye.	$P = 1.27$ D

Check: The units work out properly. To correct farsightedness, we need glasses that provide more convergence and have a positive power.

Taking It Further: What is the angular magnification, M, of these glasses?

Try It Yourself #6

An astronomical telescope with a magnifying power of 20 consists of two converging lenses, the objective and the eyepiece, located 1.00 m apart. What is the focal length of each lens?

Picture: In an astronomical telescope the two interior focal points coincide, and the magnification is the ratio of the focal lengths.

Solve:

Sketch the telescope.	
Relate *algebraically* the separation of the lenses to the focal lengths.	
Relate *algebraically* the magnification to the focal lengths.	
Substitute this last expression into that from the second step to solve for the eyepiece lens focal length.	$f_e = 0.0476$ m
Substitute this last result into the magnification expression to find the objective lens focal length.	$f_o = 0.952$ m

Check: The units check out, and the total length is in fact 1.00 m.

Taking It Further: If this telescope is used to view Jupiter when that planet is at its closest approach to Earth, what is the diameter of the image of Jupiter formed by the *objective*? Jupiter's diameter is 143 000 km, and is located 630 000 000 km away.

QUIZ

1. TRUE or FALSE: A perfectly spherical reflecting surface is free of chromatic aberrations.

2. TRUE or FALSE: A virtual image cannot be seen but must be projected on a screen.

3. For a certain lens in air, both radii of curvature are positive. Is it a converging lens or a diverging lens? Or do you need additional information to tell?

4. As we have described the astronomical telescope, the final image it forms is virtual. How then is it possible to take telescopic (astronomical) photographs by allowing the image to fall on photographic film?

5. What is the minimum height that a plane mirror must have for you to be able to see your whole body reflected in it?

6. When you place a bright light source 36.0 cm to the left of a lens, you obtain an upright image 14.0 cm from the lens, and also a faint inverted image 13.8 cm to the left of the lens that is due to reflection from the front surface of the lens. When the lens is turned around, a faint inverted image is 25.7 cm to the left of the lens. What is the index of refraction of the material?

7. A compound microscope has a tube length of 20.0 cm and an objective lens of focal length 8.00 mm. (a) If it is to have a magnifying power of 200, what should be the focal length of the eyepiece? (b) If the final image is viewed at infinity, how far from the objective should the object be placed?

Chapter 33

Interference and Diffraction

33.1 Phase Difference and Coherence

In a Nutshell

Light is an electromagnetic wave. When two or more light waves superpose (overlap), the electric and the magnetic field vectors add in accordance with the principle of superposition (vector addition). Much of this chapter deals with the distribution of light intensity when the wavefronts are partially blocked and when two or more waves superpose. The frequency of visible electromagnetic radiation—light—is between 4.3×10^{14} Hz and 7.5×10^{14} Hz. Neither our eyes nor most other instruments respond rapidly enough to make observations at such frequencies, so we will discuss only observations of light averaged over many cycles.

When harmonic waves of the same polarization, frequency, and wavelength superpose at a point, the resulting wave, which has the same frequency and wavelength as the original waves, has an amplitude that depends on the phase difference between the component waves. If the component waves are **in phase** (or equivalently, out of phase by any integer multiple of 360° or 2π radians), they **interfere constructively**, that is, the amplitude of the resultant wave is the sum of the amplitudes of the component waves. If they are 180° or π **out of phase**, they **interfere destructively**, that is, the amplitude of the resultant wave is the difference of the amplitudes of the component waves.

Suppose light takes times t_1 and t_2 to travel from a point source to point P along paths 1 and 2, respectively. If the wave source is harmonic with frequency f, at P the **phase difference** δ between the two wavefronts is $2\pi f(t_2 - t_1)$ and when the wave speed v is the same on both paths, then $t_2 - t_1 = \Delta r/v$, where Δr is the difference in the path lengths. Thus, the phase difference due to a difference in path length is $\delta = \dfrac{\Delta r}{\lambda} 2\pi = \dfrac{\Delta r}{\lambda} 360°$.

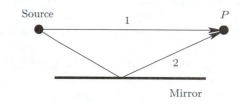

Additional phase differences are sometimes produced when waves reflect at a boundary between two transparent media. When light incident normal to the boundary reflects at such a boundary, if the wave is incident from the medium with the higher wave speed (smaller index of refraction), the reflected wave is 180° out of phase with the incident wave. If the wave is incident from the medium with the lower wave speed (higher index of refraction), there is no phase change on reflection.

If two waves differ in phase by a constant amount, they are said to be **coherent**. Coherence in optics is usually achieved by dividing and then recombining light from a single source.

Important Derived Results

Phase difference due to path-length difference $\qquad \delta = \dfrac{\Delta r}{\lambda} 2\pi = \dfrac{\Delta r}{\lambda} 360°$

Common Pitfalls

➢ Remember that a difference in path length is only one factor in determining the phase difference between light waves that have traveled from a source by two different paths. One or both of the two waves may have experienced additional phase changes due to reflection.

➢ When computing phase difference, you must divide the path-length difference by the wavelength *in the material where the path difference occurs.*

1. TRUE or FALSE: When light traveling in a transparent medium is reflected at a boundary with another medium having a higher index of refraction, the light wave undergoes a 180° phase change on reflection at the boundary

2. The main reason that some fairly smart guys (like Newton) thought that light is a propagation of particles was that diffraction and interference of light had not been observed. Why hadn't it?

33.2   Interference in Thin Films

In a Nutshell

We observe interference in the light reflected at the two surfaces of a thin film of transparent material. If the phase difference between the two reflected waves, one reflected off the front surface, the other off the back surface, is an even integral multiple of 180°, the interference is **constructive** and the intensity of the reflected light is a maximum; and if the phase difference is an odd integral multiple of 180°, the interference is **destructive** and the reflected intensity is a minimum. Mathematically, this can be expressed by the following general relationships for the phase difference:

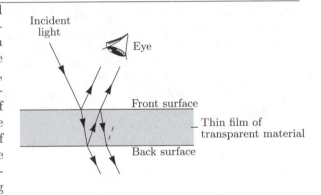

$$\delta = 2\pi m, \quad m = 0, \pm 1, \pm 2, \ldots \quad \text{constructive}$$
$$\delta = 2\pi \left(m + \tfrac{1}{2} \right), \quad m = 0, \pm 1, \pm 2, \ldots \quad \text{destructive}$$

One source of phase difference is the different path-lengths two rays of light travel. A path-length difference arises in the above example because the light reflected at the back surface travels farther than the light reflected at the front surface. For normal incidence the light reflected at the back surface travels the thickness of the film twice. This additional path length occurs within the film, so the wavelength of the light *in the film* must be used to calculate the associated phase difference.

If the thickness of the film varies, the phase difference between the two reflected waves varies accordingly. If the thickness varies over several wavelengths, **interference fringes** (maxima and minima) can be seen in the reflected light. If white light is incident on the film, the interference maxima occur at different thicknesses for different wavelengths, and we see various colors in the reflected light.

Important Derived Results

General interference conditions
$$\delta = 2\pi m \quad m = 0, \pm 1, \pm 2, \ldots \quad \text{constructive}$$
$$\delta = 2\pi \left(m + \tfrac{1}{2}\right) \quad m = 0, \pm 1, \pm 2, \ldots \quad \text{destructive}$$

Common Pitfalls

> ➤ When computing phase difference, you must divide the path-length difference by the wavelength in the material where the path difference occurs.

3. TRUE or FALSE: Monochromatic light rays reflected from the two surfaces of a film that is exactly one-half-wavelength thick always exhibit destructive interference.

4. What is the origin of the colors we see in the sunlight reflected from thin films? Explain.

Try It Yourself #1

A glass lens with index of refraction $n = 1.57$ is coated with a thin layer of transparent material of index $n = 2.10$. If white light strikes the lens at near-normal incidence, light of wavelengths 495 nm and 660 nm is absent from the reflected light. If the coating is the thinnest possible that meets the conditions given, what is the thickness of the layer?

Picture: If both wavelengths are absent, then both wavelengths experience destructive interference.

Solve:

Draw a sketch of the situation, including the two paths of light that result in the interference pattern.	
Find an *algebraic* expression for the path length difference required for destructive interference.	

Relate *algebraically* the wavelength in the film to the wavelength in air.	
Substitute this expression into that obtained in the second step to get the condition for destructive interference.	
This expression must be true for both wavelengths given in the problem. Let $\lambda_a = 660$ nm and $\lambda_b = 495$ nm. Write this expression *algebraically* for both wavelengths. They will likely be satisfied with different values of the integer m.	
Taking the ratio of the two expressions, we can find the smallest possible values for m_a and m_b, which will also correspond to the smallest possible thickness.	$t = 471$ nm

Check: You should expect thicknesses on the order of the wavelength of light, so this seems reasonable.

Taking It Further: What wavelengths are absent from the transmitted light?

Try It Yourself #2

A wedge-shaped air film is made by placing a small slip of paper between the edges of two glass flats 12.5 cm long. Light of wavelength 600 nm is incident normally on the glass plates. If interference fringes with a spacing of 0.200 mm are observed along the plate, how thick is the paper? This form of interferometry is a very practical way of measuring small thicknesses.

Picture: Interference fringes occur because the thickness of the air gap between the plates varies along the plates.

Solve:

Make a sketch of the wedge, showing two paths of light: one in which the light is reflected off the back surface of the top piece of glass, and the other in which light is reflected off the top surface of the lower piece of glass. Because the gap thickness changes with horizontal distance x, interference fringes will occur. The light paths through all the rest of the glass remains constant for all values of x.	
Determine the path-length difference required for a bright interference fringe.	
Relate the thickness of the air wedge D to the distance x traveled from the point where the two glass plates touch.	
Substitute the expression for D into the expression obtained in the second step. Knowing the separation of the fringes, solve for the thickness of the paper. Use two separate expressions for D: one for the mth fringe at a position x, and one for the $(m+1)$st fringe at a position $x+\Delta x$. This will enable you to solve for the thickness of the paper.	$t = 188\ \mu\mathrm{m}$

Check: This seems like a reasonable thickness, which would result in approximately five sheets per millimeter.

Taking It Further: If 400 nm light is used, how will the interference fringes change, if at all?

33.3 Two-Slit Interference Pattern

In a Nutshell

Young's two-slit interference experiment demonstrates conclusively the wave aspect of light. In this experiment two narrow parallel slits S_1 and S_2 in an opaque screen are illuminated by light from a single source. The light passing through each slit spreads out, and the intensity on the screen displays an interference pattern determined by the difference in path length from each slit to each point on the screen. The scale of the figure is highly distorted in that both the widths of the slits and the distance d between the slits are actually extremely small compared with the distance L from the slits to the screen. When $\ell_1 = \ell_2$ the slits S_1 and S_2 can themselves be considered to be coherent sources in phase with each other. When L is large compared to the other distances, the path-length difference between the waves can be shown to be $\Delta r = d \sin \theta$. As there are no phase changes on reflection, constructive interference occurs at angles where the difference in the path length Δr from the two slits is an integer number m of wavelengths. The integer m is called the **order number.**

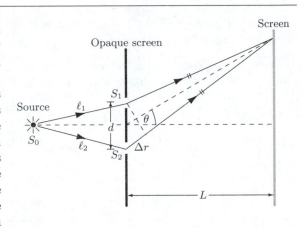

The **intensity** of any wave is proportional to the square of its amplitude. The intensity at some point in an interference pattern is thus proportional to the square of the net amplitude of the resultant wave produced by the two individual waves. The intensity of any pattern involving the interference of two, in-phase, coherent light sources of equal amplitude can be shown to be $I = 4I_0 \cos^2 \left(\frac{1}{2} \delta \right)$, where δ is the phase difference of the paths of light from the two sources, and I_0 is the intensity from either source separately.

Important Derived Results

Path difference due to two-slit interference

$$\delta = \frac{\Delta r}{\lambda} 2\pi = \frac{d \sin \theta}{\lambda} 2\pi$$

Intensity in terms of phase difference

$$I = 4I_0 \cos^2 \left(\frac{1}{2} \delta \right)$$

Common Pitfalls

> In interference and diffraction calculations, it is easy to confuse phase differences with angles between directions in space. Both quantities are expressed in angle measure, but they are quantities of completely different kinds. Adding to the confusion is that on phasor diagrams phase differences are represented as angles between directions.

5. TRUE or FALSE: Complete destructive interference of two light waves requires that they be of equal intensity.

6. The maximum intensity of light in the interference pattern produced by two narrow slits is four times, not twice, the intensity that we would see from one narrow slit alone. Where does the extra energy come from?

Try It Yourself #3

In a classroom demonstration of Young's experiment, helium-neon laser light of wavelength 632.8 nm passes through two parallel slits and falls on a screen 10.5 m away. Interference maxima on the screen are spaced 4.00 cm apart. What is the separation of the slits?

Picture: Determine the condition for constructive interference at the first maximum beyond the central peak.

Solve:

Sketch the situation. Include the two different paths of light and the angle the paths make with the centerline between the slits. Identify the path difference.	
Determine *algebraically* the tangent of the angle.	

Determine *algebraically* the condition for constructive interference at some point a distance y from the bisector of the slits on the screen.	
Because the height y at which the first maximum is observed is much smaller than the distance L, the small-angle approximation can be used. Relate the sine of the angle to the tangent of the angle.	
Substitute the expression of the previous step into the condition for constructive interference, and solve *algebraically* for the location of the mth maxima.	
We know the spacing of subsequent maxima. Use this to solve for d, by finding $\Delta y = y_{m+1} - y_m$.	
	$d = 166\ \mu\text{m}$

Check: The units work out, and this slit separation seems reasonable.

Taking It Further: The instructor slips a piece of clear plastic film 0.1 mil (2.54 μm) thick, with $n = 1.561$, over one of the slits. How does the interference pattern change?

33.4   Diffraction Pattern of a Single Slit

In a Nutshell

According to Huygens's principle, light passing through a single slit can be thought of as light emitted from several coherent point sources, rather than light from a single source. As a result, light traveling through a small slit can actually interfere with itself. This "self-interference" is called **diffraction**.

Whenever part of a wavefront is limited or blocked by an obstacle, the part or parts of the wavefront that are not blocked spread out in the region beyond the obstacle. Because of this diffraction, light passing through a single slit in an opaque barrier spreads outside the geometrical shadow of the aperture. On the figure, only the rays diffracted at a specified angle are shown. The spreading angle is approximately equal to the ratio of the wavelength and the slit width. The larger this ratio (the smaller the slit), the greater the spreading out (diffraction). The intensity pattern far away from the slit, in the limit of Fraunhofer diffraction, can be derived from Huygens's principle. For Fraunhofer diffraction from a slit of width a, the diffraction *minima* occur when $a \sin \theta = m\lambda$.

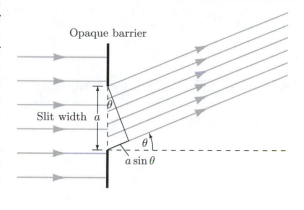

The actual pattern produced in the two-slit experiment is a combination of the two-slit interference pattern and the diffraction pattern of the light passing through each slit. As a result, when $d/a = m$, the mth interference fringe does not appear. This is because at the position defined by the location of the mth interference fringe there is a diffraction minimum from the two slits. Because of diffraction, no light leaves either slit at that angle, so there is no light present on the screen at that angle to interfere constructively.

Important Derived Results

Single-slit diffraction minima $a \sin \theta = m\lambda$, where $m = 1, 2, 3, \ldots$

Common Pitfalls

> ➤ When thinking about how Huygens's point sources combine to form interference minima, remember always to divide the slit into an even number of segments.

7. TRUE or FALSE: Diffraction of light is important only for apertures and obstructions that are very large compared with the wavelength.

8. Is there a maximum wavelength for which both the maxima and minima of the diffraction pattern from a single slit can be observed? Is there a minimum wavelength? Explain.

Try It Yourself #4

A two-slit interference pattern using 460-nm blue light is thrown on a screen 5.00 m from the slits; bright interference fringes spaced 5.00 cm apart are observed. The fourth maximum in each direction from the central maximum is missing from the pattern. What are the dimensions of the slits?

Picture: The fringe spacing will provide the slit separation as in the previous example. The fourth fringe is missing because it coincides with the first minimum in the diffraction pattern of a single slit. This fact can be used to determine the width of the slits.

Solve:

Sketch the situation. Include the different paths of light and the angle the paths make with the centerline between the slits.	
Determine *algebraically* the tangent of the angle that light from each slit makes with respect to the centerline between the slits.	
Determine *algebraically* the condition for constructive interference at the screen.	
Apply the small-angle approximation, and use it to solve *algebraically* for y_m, the positions at which interference maxima occur.	
Use the spacing between subsequent maxima to find the slit separation d.	$d = 46.0 \ \mu\mathrm{m}$
Find an *algebraic* expression for the location of the missing fourth interference maximum.	

Determine *algebraically* the position of the first diffraction minimum from a single slit and apply the small-angle approximation.	
Because the fourth interference maximum coincides with the first diffraction minimum, we can set the expressions for y in the two previous steps equal to each other to solve for the slit width a.	$a = 11.5 \ \mu\text{m}$.

Check: This seems reasonable. The slit is larger than the wavelength.

Taking It Further: If the slit widths are slowly increased in size, eventually a different interference maximum will disappear. Will it be the third or fifth maximum? Explain.

Try It Yourself #5

In a lecture demonstration of diffraction, a laser beam of wavelength 632.8 nm passes through a vertical slit 0.250 mm wide and strikes a screen 12.0 m away. How wide is the central maximum (the distance between the first diffraction minima on each side of the central maximum) of the diffraction pattern on the screen?

Picture: Find the distance of the first diffraction minimum from the central maximum and multiply by two.

Solve:

Sketch the situation.	

Determine *algebraically* the position of the first diffraction maximum from a single slit.	
Apply the small angle approximation and solve *algebraically* for y, the position of the first minimum.	
Multiply the position arrived at in the previous step by 2 to find the width of the central maximum.	$\Delta y = 6.07$ cm

Check: The units work out, and this is a respectably sized spot.

Taking It Further: If the slit is illuminated with light of a larger wavelength, how will the width of the central maximum change, if at all? Explain.

33.5   Using Phasors to Add Harmonic Waves*

In a Nutshell

The superposition of two harmonic waves of the same frequency and wavelength at a point results in a harmonic wave of the same frequency and wavelength. The resultant wave can be found by a geometric construction that is equivalent to ordinary vector addition, as shown. Each component wave is represented by a vector (a phasor) whose length is the amplitude of that wave and whose direction is determined by its phase angle. Thus, the phase difference δ in the direction of two phasors A_1 and A_2 is the phase difference between the two waves. If the two waves are coherent, that is, if the difference in their phases is constant, the angle between the directions of the two phasors remains constant. Adding two or more phasors vectorally produces a resultant phasor whose length and direction represent the amplitude and phase of the resultant wave.

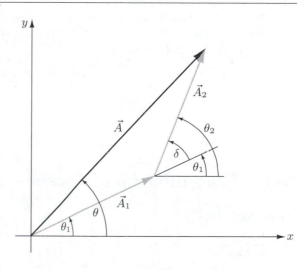

Using phasors, the intensity of the single-slit diffraction pattern can be shown to be $I = I_0 \left(\dfrac{\sin \frac{1}{2}\phi}{\frac{1}{2}\phi} \right)^2$, where the phase difference ϕ is given by $\phi = (2\pi/\lambda) a \sin \theta$ and a is the width of the slit.

The intensity of the double-slit pattern, which combines both the interference between the slits and the diffraction of the individual slits can be shown to be $I = 4I_0 \left(\dfrac{\sin \frac{1}{2}\phi}{\frac{1}{2}\phi} \right)^2 \cos^2 \frac{1}{2}\delta$, where ϕ is the same as for the single-slit pattern and $\delta = (2\pi/\lambda) d \sin \theta$, where d is the separation of the slits.

As more coherent sources with a fixed spacing are added to a system, the locations of the principal interference maxima remain fixed. However, the intensity and sharpness (narrowness) of the principal maxima increase as the number of sources increases. In addition, for N total coherent sources, $N-2$ less intense secondary maxima appear equally spaced between the primary maxima.

Important Derived Results

Single-slit diffraction intensity pattern

$$I = I_0 \left(\frac{\sin \frac{1}{2}\phi}{\frac{1}{2}\phi} \right)^2$$

Interference-diffraction intensity for two slits

$$I = 4I_0 \left(\frac{\sin \frac{1}{2}\phi}{\frac{1}{2}\phi} \right)^2 \cos^2 \frac{1}{2}\delta$$

Common Pitfalls

➤ When interference is due to three or more equally spaced coherent sources or slits, the location on the screen of the principal interference maxima depends only on the spacing of the sources, not on how many of them there are. Both the sharpness of the principal maxima and the number of secondary maxima between them increase with the total number of sources or slits.

*Optional material

9. TRUE or FALSE: The principal interference maxima produced by light from four identical slits in an opaque screen a distance d apart occur at angles that are twice those of the interference maxima that would be produced by light from the middle two slits.

10. Sketch the intensity pattern you expect to see from six equally spaced coherent sources separated by a distance d. Your vertical axis should be intensity, and your horizontal axis should be $\sin\theta$. The sketch should include at least three primary maxima; label their locations.

33.6   Fraunhofer and Fresnel Diffraction

In a Nutshell

The diffraction pattern observed at points for which the rays from an aperture or an obstacle are nearly parallel (typically the pattern is observed at large distances from an aperture) is called a **Fraunhofer diffraction** pattern. The diffraction pattern observed very close to an aperture or an obstacle, when the rays from that object can no longer be approximated as parallel, is called a **Fresnel diffraction** pattern. These two patterns, from the same aperture or object, are quite different from each other, as illustrated on pages 1159 and 1160 in the text.

33.7   Diffraction and Resolution

In a Nutshell

For any optical instrument in which light enters through a finite aperture, diffraction limits the instrument's resolution. The angle θ at which the first minimum in the Fraunhofer diffraction pattern of the light from a circular aperture of diameter D occurs is $\sin\theta = 1.22\lambda/D$, as shown in Figure 33-34 on page 1160 of the text. In the small-angle approximation, $D \gg \lambda$, this becomes $\theta \approx 1.22\lambda/D$. This angle is the smallest angle that can be resolved, and is known as the critical angular separation, α_c. The derivation of this result is beyond the scope of both the text and this study guide. The images of objects separated by angles smaller than this are not distinguishable because they overlap. This is known as the **Rayleigh criterion** of resolution. An astronomical instrument must have a large aperture for the sake of resolution as well as for light-gathering power.

Important Derived Results

First minimum, circular aperture diffraction $\sin\theta = 1.22\dfrac{\lambda}{D}$

Common Pitfalls

➤ When determining the minimum angular resolution of an instrument, remember to use the diameter, not the radius, of the aperture.

11. TRUE or FALSE: Diffraction imposes a fundamental limitation on the resolving power of telescopes and microscopes.

12. Under the most optimistic atmospheric conditions at the premium site for land-based observing (Mauna Kea, Hawaii, elevation ≈ 4.27 km), an optical telescope can resolve celestial objects that are separated by one-fourth of a second of arc (arc-sec). The viewing never gets any better than this because of atmospheric turbulence, which makes the images jitter. What minimum diameter aperture is necessary to provide arc-sec resolution due to diffraction? Is there ever any point in building a telescope much bigger than this? Explain.

Try It Yourself #6

Two point sources of light of wavelength 500 nm are photographed from a distance of 100 m using a camera with a 50.0-mm focal length lens. The camera aperture is 1.05 cm in diameter. What is the minimum separation of the two sources of they are to be resolved in the photograph, assuming the resolution is diffraction limited?

Picture: The minimum resolution is determined by Rayleigh's criterion. The limiting factor is the size of the aperture. This angular separation corresponds to a linear separation 100 m away.

Solve:

Draw a sketch of the physical situation.	
Write *algebraically* an expression for the angular position of the first diffraction minimum for this circular lens, using the small angle approximation.	
The critical angular separation requires that the central diffraction maximum of one source be located at the first minimum of the diffraction pattern of the other source (Rayleigh's criterion). Use this fact to find an *algebraic* expression for the minimum angular separation of the sources.	

Because of the small-angle approximation, the critical angle is approximately equal to the tangent of the critical angle. Use this to solve for the source separation distance.	
	$d = 5.81$ mm

Check: This seems like a reasonable spatial resolution.

Taking It Further: What is the separation of the two images when the sources are at this minimum separation? Explain.

Try It Yourself #7

The world's largest refracting telescope is at Yerkes Observatory in Williams Bay, WI, less than seven miles from the house where I grew up. Its objective is 1.02 m in diameter. Suppose you could mount the telescope on a spy satellite 200 km above the ground. Assuming that the resolution is diffraction limited, what minimum separation of two objects on the ground could it resolve? Take 550 nm as a representative wavelength for visible light.

Picture: Use Rayleigh's criterion to find the minimum angular separation for diffraction-limited resolution. Find the linear separation for this angle.

Solve:

Draw a sketch of the physical situation.	
Determine the angular position of the first diffraction minimum for this circular lens, using the small-angle approximation.	

The critical angular separation requires that the central diffraction maximum of one source is located at the first minimum of the diffraction pattern of the other source.	
Because of the small-angle approximation, the critical angle is approximately equal to the tangent of the critical angle. Use this to solve for the source separation distance.	$d = 13.1$ cm

Check: Although this telescope couldn't distinguish facial features, it could certainly count the number of people in a crowd!

Taking It Further: Because of atmospheric turbulence, objects on the surface of Earth can be distinguished only if their angular separation is at least 1.00 arc-sec. How far apart would two objects on Earth's surface be if they subtend an angle of 1.00 arc-sec? Compare this with your answer above.

33.8 Diffraction Gratings*

In a Nutshell

A **diffraction grating** consists of a large number of equally spaced lines or slits. The interference maxima in light that has passed through the grating can be extremely sharp. A diffraction grating is used to analyze the spectrum, the distribution of energy among different wavelengths, of light from various sources. The expression for the resolving power of a grating is $R = \lambda/|\Delta\lambda| = mN$, where N is the number of slits illuminated, m is the order of the principal interference maxima, and $|\Delta\lambda|$ is the minimum difference in wavelengths that can be resolved.

Important Derived Results

Resolving power of a diffraction grating
$$R = \frac{\lambda}{|\Delta\lambda|} = mN$$

---

*Optional material

Common Pitfalls

> ➤ Do not confuse line sharpness with line spacing. A grating has high resolution if you can use it to separate cleanly two light waves of very nearly the same wavelength—that is, after passing through the grating the two waves form two distinct lines. There are two separate issues involved in this: the separation between the centers of the two lines, which is determined by the spacing between slits, and the sharpness (narrowness) of each line, which increases as the number of slits illuminated increases. The resolving power of a grating depends upon both these properties.

13. TRUE or FALSE: For a given line spacing, a larger diffraction grating is capable of better spectral resolution.

14. It often seems to me that a diffraction grating should more properly be called an interference grating. What do you think? Why?

Try It Yourself #8

A diffraction grating with 10,000 lines per centimeter is used to analyze the spectrum of mercury. Find the angular separation, in first order, of the two spectral lines of wavelength 577 and 579 nm.

Picture: The angles of the spectrum are the angle for the first-order primary interference maxima from the grating and depend on the line spacing of the grating.

Solve:

Draw a sketch of the physical situation.	
Determine *algebraically* the condition for the principal interference maxima.	

The first-order maxima require that $m = 1$. Use this to solve for the angular position of the first interference maximum for each of the two wavelengths.	
Determine the angular separation.	$\Delta\theta = 0.14°$

Check: While correct, this is not a very large angular separation.

Taking It Further: How wide must the beam of light be on the grating in order to resolve these two spectral lines? Explain.

QUIZ

1. TRUE or FALSE: Light from two coherent sources that are not in phase does not produce an interference pattern.

2. TRUE or FALSE: Light passing through two narrow slits in a barrier exhibits both interference and diffraction.

3. Antireflecting (AR) coatings for lenses are made with a thickness equal to one-fourth of the wavelength of the light. Is this one-fourth of the wavelength in air, in the coating, or in the glass lens? Is the index of refraction of the coating greater or less than that of glass? Why?

4. What is the advantage of the electron microscope over a light microscope? Explain.

5. The apparatus sketched in the figure is known as Fresnel's double mirror. Like Lloyd's mirror, it is a means of realizing two coherent light sources without using slits. How does it work?

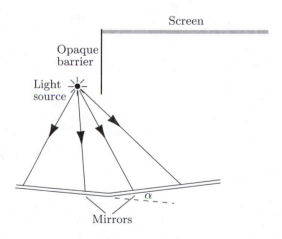

6. Light of wavelength 500 nm falls at near normal incidence on two flat glass plates 8.00 cm long that are separated at one end by a wire with a diameter as shown. Find the spacing of the dark interference fringes along the length of the plates.

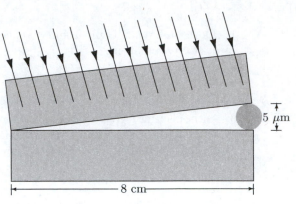

8 cm

5 μm

7. A thin film of soap solution with $n = 1.33$ has air on either side and is illuminated normally with white light. Interference minima are visible in the reflected light only at wavelengths of 400, 480, and 600 nm. What is the minimum thickness of the film?

Answers to Problems

Chapter 21

COMMON PITFALLS

1. False. "Conservation of charge" refers to the fact that the total charge of the universe is constant.

2. Our current understanding is that like charges repel and opposite charges attract. So if a charged insulating object either repels or attracts both the charged glass (positive) and charged rubber (negative) rods, then you might have a new kind of charge. However, you also might just be witnessing a phenomenon called *polarization* (see Section 21-2).

3. False. Induction always results in an attractive force.

4. When the plastic comb is run through your hair, electrons are transferred from your hair to it, so the comb acquires a net negative charge. The electric field of the comb polarizes (that is, induces an electric dipole moment in) the bit of paper along the field direction. This results in a positive charge on the edge of the paper nearest the comb and a negative charge of equal magnitude on the edge farthest from the comb. Because the electric field of the comb is non-uniform, it attracts the nearby positive charge more strongly than it repels the more distant negative charge. When the bit of paper actually touches the comb, some negative charge is transferred from the comb to the paper, giving it a net negative charge. Because like charges repel, the bit of paper is now repelled by the comb.

5. True. $\vec{F}_{12} = -\vec{F}_{21}$ because $\hat{r}_{12} = -\hat{r}_{21}$.

6. The main reason is humidity. The higher the humidity, the greater the rate at which static charges leak off charged objects into the air.

7. False. The magnitude decreases with the cube of the distance from the dipole.

8. In electrostatics the field is just a computational device, and using it is merely a matter of convenience. However, if the charges creating the field move, the fact that the field propagates at the speed of light also allows us to understand that changes in the field, and hence force on other charged objects don't change immediately. If you continue your studies in physics, you will see that in electrodynamics the field is necessary for energy and momentum to be conserved.

9. False. The electric field of a dipole is the superposition of the electric fields of a negatively charged particle and a positively charged particle., which is not parallel to the dipole moment.

10. For every two electric field lines pointing toward the −2 pC charge, there should be three electric field lines leaving the +3 pC charge.

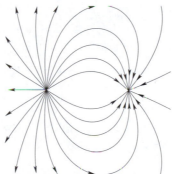

11. True. In a uniform electric field the two forces acting on the dipole are equal in magnitude and opposite in direction. However, a uniform electric field may exert a torque on the dipole.

12. No. The electric field lines point in the direction of the electric force, but the force is in the direction of the acceleration, not the velocity.

TRY IT YOURSELF–TAKING IT FURTHER

3. If both charges were negative, they would repel. As a result the force on q_2 would be in the $-\hat{\imath}$ and $-\hat{\jmath}$ directions.

4. Yes, at the same location as the +4.00 μC charge. The force from the negative charge will now be repulsive and that from the positive charge will be attractive, but their magnitudes will be equal, so the net force will still be zero.

5. The electric field due to the other charges will remain the same, so just use $\vec{F} = q\vec{E}$, substituting the value of the new charge.

6. In order for the electric field at point P to change values, at least one of the charges that create that field would have to move. We know there must be at least one charge involved, even though we are

not told anything about the configuration of charges that actually create the field. This is one of the advantages of the electric-field concept. We don't need to know the details of the charge distribution creating the field, just the value of the field.

7. Even though the initial speed of the electron is 2.510^7 m/s, its kinetic energy is so small because the mass of an electron is also quite small. This is some of the reason that energy units of electron volts (eV) are often used instead of joules when talking about individual particles rather than macroscopic boxes. Note that this is a relativistic speed, so to do the calculation properly requires special relativity. See Chapter R for more information.

8. The electron accelerates in the $-\hat{j}$ direction. Because the electron has less mass, the magnitude of its acceleration is larger. As a result, the y component of its velocity will be greater, and it will be deflected further, but in the $-y$ direction.

QUIZ

1. False

2. False

3. When the plastic comb is run through your hair, electrons are transferred from your hair to it, so the comb acquires a net negative charge. As shown in Figure (a), this charge repels the free electrons in the aluminum, which results in some of the free electrons moving to the part of the can away from the comb. If the net charge on the can is zero, this leaves the part of the can next to the comb positively charged (with an electron deficit) and the part of the can away from the comb negatively charged (with an electron excess). This results in a net attractive force exerted by the comb on the can because the magnitude of the attractive force exerted on the nearby positive charge exceeds the magnitude of the repulsive force exerted on the more distant negative charge. When the comb touches the can, some electrons are transferred from the comb to the can, giving the can a net negative charge. When the negatively charged comb is again brought near the can, the force is again attractive, even though the net charge on the can is negative, as shown in Figure (b), and the attractive force on this positive charge exceeds the repulsive force on the larger, but considerably more distant, negative charge.

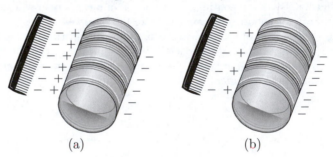

(a)   (b)

4. The two charges must have the same sign, and $|q_2| > |q_1|$.

5. When the suspended object is attracted to the glass rod, it is not necessarily charged at all. However, when the object is repelled, it must have the same charge as the glass rod.

6. $E = 1706$ N/C; $t = 3.7310^{-11}$ s

7. (a) $E_x = (-2kq/x^2 + kq/(a-x)^2)\hat{i}$; (b) $E_x = (2kq/x^2 + kq/(a-x)^2)\hat{i}$; (c) $E_x = (2kq/x^2 - kq/(x-a)^2)\hat{i}$; (d) $x = a(2 + \sqrt{2})$

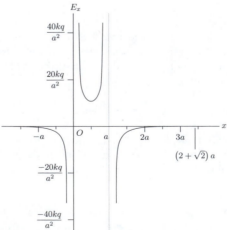

Chapter 22

COMMON PITFALLS

1. False. Near a uniformly charged plane, the magnitude of the electric field does not vary with the distance from the plane.

2. Electric fields obey the law of superposition—that is, to find the field at some point P due to a distribution of charges we simply add up (vectorially) the electric field due to each individual charge. We know from calculus that an integral is simply a sum of very small, continuously distributed quantities.

3. False. Electric flux is calculated via a dot product, so it is a scalar. Electric flux can be positive or negative, however. The sign of the flux indicates the direction of the electric field lines, which is a vector quantity.

4. The electric field $\vec{E}$ in Gauss's law is the total electric field due to all charges both inside and outside the surface. However, the results are unaffected if $\vec{E}$ represents the electric field due only to the charges inside the surface. The net contribution to the electric flux out of a surface due to any charge located outside that surface is zero.

5. True

6. The electric field a distance r from the center of a uniformly charged sphere is kQ/r^2, where Q is the charge inside a sphere of radius r. Inside the sphere, as one moves out from the center Q increases with the cube of the distance from the center of the sphere. Thus the electric field increases as r rather than decreases as $1/r^2$.

7. True

8. As shown in Figure 22-32 on page 751 of the text, just outside the surface of a conductor the electric field is the superposition of two electric fields, one due to the nearby surface charge and the other due to all other charges. These two fields cancel just inside the surface of the conductor so they must have equal magnitudes. Just outside the surface the two fields are in the same direction so the net electric field is twice as large as the electric field due to the nearby surface charge alone.

TRY IT YOURSELF–TAKING IT FURTHER

1. If all the charge on the line were located at the line's center, the electric field would be $\vec{E} = (-3070 \text{ N/C})\hat{\imath} + (+4600 \text{ N/C})\hat{\jmath}$, which is 2–3 percent different from the actual value.

2. The charge distribution is still symmetric about the y axis, so $E_x = 0$ still. However, the lower portion of the disk will have a positive charge equal in magnitude to the negative charge on the upper half of the disk. This positive charge will be closer to the point P, so its electric field will dominate. As a result the field at P will now point in the $-y$ direction.

3. The net flux would now be negative because the prism now encloses a net negative charge which contributes to the electric field on each face of the prism.

4. The flux through the sphere would still be zero, as the sphere does not enclose any net charge.

5. The exact same symmetry arguments and the exact same algebraic expression resulted for the electric flux through the Gaussian cylinder in each step. This makes sense because the cylindrical symmetry is identical in each case. The only difference in calculating the electric field in each region of space is the expression for the enclosed charge.

6. You get the same answer. This is because outside a spherical charge distribution the electric field is indistinguishable from the electric field that would be created if all the charge were located at the center of that charge distribution.

7. The problem would not change at all, since for both the shell and the solid conductor the excess charge will sit on the outside surface.

8. If the shell were insulating, the charges would not be free to move around. As a result, we would not be able to calculate the surface charges in the same way at all. There will still be some surface charge layer due to polarization of the insulator's molecules, but that is a much more challenging problem to solve.

QUIZ

1. False. A zero net electric flux out of a surface can be accomplished with a positive flux out of some parts of the surface and a negative flux out of other parts. For an example, see Figure 22-14 on page 738 of the text.

2. True

3. No. All it means is the net charge enclosed by the surface is zero. The charge can have any distribution within the surface.

4. The charge on conductors always resides on the surfaces of the conductor. The electric field inside a conductor in static equilibrium is equal to zero. Since the electric field is discontinuous by an amount

equal to σ/ϵ_0, the field outside the conductor must also have that value.

5. Gauss's law, correctly applied, does not give the same result for the two cases.

For the uniformly charged infinite plane we consider a cylindrical Gaussian surface with the ends of the cylinder both parallel to and equidistant from the plane. From the symmetry we know that on the curved side of the can the normal component of the electric field is everywhere zero, so the flux through this side is zero. Also from symmetry, we know that on the ends of the can the normal component of the electric field is uniform and equal to its magnitude. Thus the total flux out of the cylinder is $\phi = 2E_n A = 4\pi k\sigma A$ so $E_n = 2\pi k\sigma$.

For the uniformly charged disk we again consider a cylindrical Gaussian surface with the ends of the cylinder both parallel to and equidistant from the disk. There is insufficient symmetry to allow us to conclude either that on the curved side of the can the normal component of the electric field is zero or that on the ends of the can the normal component of the electric field is uniform and equal to its magnitude.

6. (a) $Q = 1.610^{-8}$ C; (b) $\vec{E} = (9.986210^{-3}$ N/C$)\hat{\jmath}$; (c) $\vec{E} = 9.984810^{-3}$ N/C$)\hat{\jmath}$; (d) $\vec{E} = 3595.0$ N/C$)\hat{\jmath}$; (e) $\vec{E} = 3594.8$ N/C$)\hat{\jmath}$.

7. $\phi_{\text{disk}} = 0$; $\phi_{\text{hemisphere}} = 2\pi a^3 \rho/(3\epsilon_0)$; $\phi_{\text{total}} = 2\pi a^3 \rho/(3\epsilon_0)$.

Chapter 23

COMMON PITFALLS

1. False. A volt is a joule per coulomb, which is energy per unit charge. Power is energy per unit of time.

2. Any negatively charged particle, including an electron, will move in a direction opposite to that of the electric field. The electric field always points in the direction of *decreasing* potential, so the electron will move in the direction of *increasing* potential. It will also always move in the direction of decreasing potential energy.

3. False. Charges of like sign repel each other. If they are released from rest, they fly apart, their kinetic energy increases, and their potential energy decreases.

4. The electric potential is associated with the location. It is the ratio of electrostatic potential energy to charge that a test charge at the location would have—if a test charge were at the location.

5. True

6. The electric field points in the direction of decreasing potential. For the two points to be at the same potential, the line joining the points must be perpendicular to the direction of the electric field.

7. True

8. If the electric field is zero, the electric potential is constant, but not necessarily zero. Mathematically this this can be understood because the field is the gradient of the potential.

9. False. Molecules are ionized when dielectric breakdown occurs, but the electric field does not ionize the molecules directly. The electric field accelerates existing ions, which, when they collide with another molecule, have enough energy to cause that second molecule to ionize.

10. Dielectric breakdown is not a phenomenon of potential. Breakdown occurs when the *field* is sufficiently large. The electric field near the surface of a conductor is larger where the conductor has a smaller radius of curvature (see page 787 of the text).

11. True

12. The factor of $\frac{1}{2}$ eliminates the double-counting of potential energy from pairs of charges. If you consider only two charges, the summation without the factor of $\frac{1}{2}$ would result in $U = (kq_1q_2/r_{12}) + (kq_2q_1/r_{21})$. These two terms are equal, but their sum is twice the electrostatic potential energy, so the factor of $\frac{1}{2}$ is required.

TRY IT YOURSELF–TAKING IT FURTHER

1. The potential difference between the two points would remain the same—it is independent of the charge. If a negative charge were placed at rest at point a it would never arrive at point b unless some external agent acted upon it. The negative charge would accelerate in the $-x$ direction, which is the direction of lower potential energy for the electron in this electric field.

2. The external agent has to do no net work. This is possible because for $x < 0$ the electric field points in the $+x$ direction, and for $x > 0$ the electric field points in the $-x$ direction. Furthermore, the magnitude of the electric field is symmetric about $x = 0$. So the field does positive work, initially increasing the speed of the test charge until it reaches $x = 0$, at which point the field does negative work, slowing the test charge until it reaches $x = +30.0$ cm.

3. If a negative charge is brought in it still experiences an identical change in electric potential because the potential is due to the other charges in the distribution. The work done by an external force will

now be $W_{\text{ext}} = -0.204$ J because the field does positive work on the negative charge.

4. The work required is simply the charge multiplied by the change in potential: $+4.2810^{-15}$ J, or 26 700 eV.

5. The sodium ion has a charge of $+2e$, so it will accelerate in the direction of the electric field found, generally in the $+x$, $-y$, and $-z$ directions.

6. Yes, the potential is zero when $\frac{2}{x} = \frac{1}{a-x}$. This is possible because the positive charge creates a positive potential and the negative charge creates a negative potential, so they can add to zero.

7. Your answers should look something like this. The solid lines are for the charged disk, and the dashed lines are for the infinite plane. The potential of the infinite plane has some arbitrary maximum value on the charged plane, and then decreases linearly, with the same slope, on either side of the charged plane.

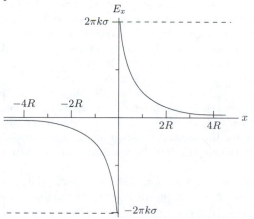
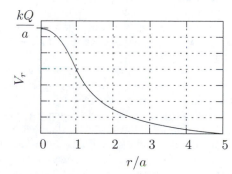

As the radius of the disk goes to infinity, the plots of the field and potential should approach those of the infinite sheet shown above.

8.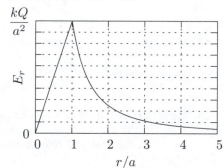

9. The minimum radius would be much smaller because the breakdown field is larger. Larger fields occur with smaller radii of curvature.

12. In either case, the charges all repel one another. The charges also have the same magnitude.

QUIZ

1. True

2. True

3. Both electric field and potential result from the existence of a charge. If a charge exists, it will create both of these physical quantities. In order to have a nonzero potential energy, a force must be able to do work on an object as it is moved. If only one charge exists, then there is no electric force acting on it from some other charge, so that force can do no work, so there is no potential energy.

4. This statement makes sense only if the zero point of the electric potential has been previously defined.

5. $\Delta U = q \, \Delta V$. In isolation an object will always accelerate toward a region of lower potential energy as required by the work–energy theorem. If ΔU is negative, and the charge q is negative, then ΔV must be positive. Another way to think about this question is to recall that electric field lines point in the direction of decreasing electric potential. Since negative charges accelerate in a direction opposite to those field lines, they must be accelerating toward a region of higher electric potential.

6. (a) $V_c - V_b = -49.9$ V, independent of the location for $V = 0$. (b) The equipotential surfaces are equidistant from the line of charge. They are circular cylinders coaxial with the line charge.

7. 71 900 V with respect to the potential at infinity.

Chapter 24

COMMON PITFALLS

1. False. Capacitance does not have dimensions of charge. It is the ratio of stored charge to the potential difference between its conductors. As the charge increases or decreases, so does the potential difference between the conductors.

2. The potential difference depends on the electric field between the conductors. This electric field is proportional to the charge on each of the plates, so the potential difference is proportional to the charge. Since Q and V appear as a ratio in the definition of capacitance, the value of Q will cancel out, leaving only geometric and material properties.

3. False. It is $\frac{1}{2}QV$.

4. The energy stored is equal to the work required to separate the charges in the capacitor. Because there is no field between the conductors initially, the first electron requires no work to move. However, once there is some charge separation, there is an electric field, and the work required to move subsequent charges increases linearly with the number of charges separated until finally the last charge requires a work of QV to be moved. Adding up all the work required results in the factor of $\frac{1}{2}$.

5. False. They have the same potential difference across the plates, but not the same charge.

6. Attaching capacitors in series allows the total potential difference to be split among the individual capacitors. If the voltage across a capacitor becomes too large, the resulting electric field will cause dielectric breakdown, ruining the capacitor. As a result, every capacitor has a maximum allowable voltage that cannot be exceeded. That is the main reason for attaching a group of capacitors in series.

7. True. For a given charge stored on the electrodes the electric field between the electrodes is smaller in the presence of a dielectric.

8. The electric field in the region between the plates is the same for both capacitors. The potential difference V across a capacitor and the electric field $\vec{E}$ in the region between the plates are related by the equation $V = \left| -\int \vec{E} \cdot d\vec{\ell} \right| = Ed$. If the potential difference across each capacitor is the same (which must be the case for the capacitors in parallel), then the electric fields in the regions between the plates must also be the same. However, A stores more charge.

TRY IT YOURSELF–TAKING IT FURTHER

1. Since charge remains constant, the surface charge density also remains constant. This means that the electric field also remains constant. The voltage between the plates, however, is reduced because the quantity d is reduced.

2. The quantity $r_2 - r_1$ in the denominator will get smaller, so the capacitance will increase, which means the capacitor will store more charge per volt. In addition, the numerator will also get larger, which will also increase the capacitance.

3. $U = Q^2/(2C)$. For an isolated capacitor Q is constant. Since $C = \epsilon_0 A/d$ for an air-filled parallel-plate capacitor, when d is doubled U is doubled in this case.

4. The capacitance will still increase by a factor of 4. The charged stored will remain constant. V is reduced by a factor of 4. The energy is reduced by a factor of 4.

5. Capacitance is charge per volt. A larger capacitance means more charge can be stored per volt of potential difference. In other words, for a given stored charge, a capacitor with a larger capacitance requires a smaller voltage than a capacitor with a smaller capacitance.

6. The total capacitance is the sum of the individual capacitances. This means that for a given voltage capacitors in parallel store more charge than any one of the individual capacitors.

7. The plates will attract each other. But it is more likely that the plates will spark across the air gap before you can pull the Pyrex out completely, thus discharging the capacitor.

8. The capacitance of that region is reduced. The series capacitance is also further reduced, so the total capacitance is reduced.

QUIZ

1. False. Capacitance is the ratio of charge to potential difference. Capacitance is independent of the charge because the electric field, and thus the potential difference, are proportional to the charge—that is, if you double the charge, the potential difference also must double, and the capacitance remains unchanged.

2. False. This is the expression for energy per unit volume.

3. If the free surface density and the bound surface charge density were equal, then their electric fields would completely cancel. However, to sustain a bound surface charge requires that the dielectric be polarized, which requires a nonzero electric field. Therefore, the bound surface charge density is always less than the free surface charge density.

4. It depends. Inserting a dielectric always increases the capacitance of a capacitor. If the capacitor is

isolated, we can use $U = Q^2/(2C)$, from which we see inserting the dielectric will cause the stored energy to decrease. However, if the capacitor is attached to a battery, the potential is constant and we can use $U = (1/2)CV^2$, and in this case the stored energy increases.

5. The larger dielectric strength increases the maximum voltage that can be applied across the electrodes. The larger dielectric constant decreases the voltage required to store a given amount of charge, thus increasing the capacitance. The dielectric material can be used to maintain the separation of the electrodes.

6. 12

7. $C_1 = \epsilon_0 A/d$, $C_2 = \epsilon_0 A/(d - d')$, independent of the location of the slab.

Chapter 25

COMMON PITFALLS

1. False. Because the free electrons carry a negative charge, the current direction is opposite to the direction of the drift velocity.

2. There is no contradiction. If a conductor is in electrostatic equilibrium, the electric field within it must be zero. However, any conductor that is carrying a current is definitely not in electrostatic equilibrium.

3. True

4. The equation for resistivity is $R = \rho L/A$. If the geometric factors L and A are the same, then if the resistance is twice as large, the resistivity is also twice as large.

5. False. This equation can be used for any resistor. For ohmic resistors R is constant. For other resistors R can vary with current and voltage.

6. The terminal voltage V_t of an *ideal* battery is constant. Therefore lowering the resistance results in an increased current I. The rate of Joule heating in the resistor is equal to the product $IV_t = V_t^2/R$, so the rate of heating will increase.

7. True. The electric field is in the direction of decreasing potential. Because the electrons carry a negative charge, the force exerted on them by the electric field is directed opposite to the electric field and in the direction of increasing potential.

8. The voltage drop across each bulb in the old series string was about 1/50 of 110 V, or 2.2 V. The modern parallel connection puts the full 110 V across each bulb. Placing 110 V across one of the old bulbs, designed to operate at 2.2 V, would result in excessive current in the filament, which would burn out the bulb at once, perhaps spectacularly.

9. False. An ammeter has a low-resistance shunt resistor in parallel with a galvanometer. An ammeter with a large resistance would significantly decrease the current in a circuit—which, of course, is undesirable if you want to know what the current is when the meter is not present.

10. If you remember that an ammeter is a low-resistance device, then you will realize that when 112 V is placed across it, the current will be very large. In this case there was a loud pop like a firecracker, some smoke, and a very startled co-worker. The 112 V delivered a mortal blow to the meter.

11. False. Kirchhoff's rules apply to all circuits except those operating at very high frequencies.

12. It cannot be changed instantaneously because the resistor in the circuit limits the current (the rate of flow of charge). Because the current is finite, it requires time for the charge to flow on and off the capacitor plates.

TRY IT YOURSELF–TAKING IT FURTHER

1. No. Since the electrons are traveling at nearly the speed of light, one electron will pass by the point on the synchrotron several times during the course of one hour. Therefore many fewer electrons are required—only about 4×10^{11}.

2. Since an electron has 1.602×10^{-19} C, this charge corresponds to 3.74×10^{17} electrons.

3. Silver has a slightly smaller resistivity, so its resistance would be slightly smaller and the current would be slightly larger. The electric field, however, would remain the same.

4. The dissipated power is proportional to V^2. The ratio of potentials is $110^2/12^2 = 84.0$, so it will take 84 times as long to heat your water with the car battery. You'll be waiting a while.

5. The load resistance can be calculated with $\mathcal{E} = I(r + R_{load})$. $R_{load} = 0.417 \; \Omega$ for the 30.0-A case and 0.134 Ω for the 80.0-A case. The emf and internal resistance of the battery are constant, so if I increases then the total resistance must decrease.

6. $P = I^2R = V^2/R = IV$. Clearly, for a given potential the larger resistor will have less power dissipated. Because current is inversely proportional to R, if R increases then I will be reduced. But I appears as the square, so less power will again be dissipated.

7. The current is larger. The 35-Ω resistor is in series with the 65-Ω resistor, so the total resistance

of that branch is 100 Ω. The 60-Ω resistor by itself is in parallel with that branch. To obtain the same potential difference across both branches requires that the current in the smaller-resistance 60-Ω branch be larger.

8. If this battery is rechargeable, it is being recharged, rather than providing charge to the circuit. If the battery is not rechargeable, then it is simply heating up, and may explode.

9. A negative current simply means that your initial guess for the direction of positive current was incorrect. The current in that segment is actually in the opposite direction of your initial guess.

10. The ammeter must be inline, somewhere in the branch containing I. The voltmeter must be attached parallel to R_1, as in the figure below.

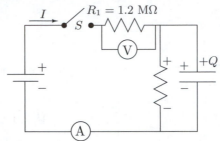

11. The ammeter must be inline, somewhere in the branch containing R_2. The voltmeter must be attached parallel to C, as in the figure below.

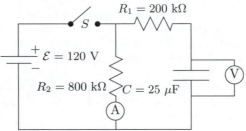

QUIZ

1. False. The drift velocity is the average velocity of the free electrons, not the average speed. When the current is zero, the average velocity of the free electrons is zero; but at room temperature the average speed is very large.

2. False. The time constant is the product RC. Check the units using the definitions of resistance ($R = V/I$) and capacitance ($C = Q/V$).

3. This depends on what you mean. The time required to charge the capacitor to a given fraction (say, 99 percent) of its final charge depends only on the time constant RC. However, the time to charge the capacitor to, say, 5 V with a 6-V battery is longer than the time to charge it to 5 V with a 12-V battery.

4. The energy is dissipated, heating up the wire. If an electron collides with the ion lattice, it transfers energy to the lattice. In this way the work done by the electric field on the electrons shows up as an increase in the thermal energy of the lattice.

5. Just calculating the resistance of the filament $R = V/I$ gives 190 Ω when 120 V is applied and 35 Ω when 3 V is applied. Noting that the resistance varies with the current might lead you to conclude that the filament is made of a nonohmic material. This conclusion would not be justified. The temperature of the filament with 120 V across it is much greater than its temperature with only 3 V across it. Most lamp filaments are made from the metal tungsten. Like most metals, tungsten is ohmic, with a resistivity that increases with temperature. The operating temperature of a typical light bulb filament is about 2800 K, so we expect the resistance of the filament to be higher when 120 V are across it even though tungsten is ohmic. If the temperature of a tungsten filament were kept constant, its resistance would be independent of current.

6. (a) $R = 0.096$ Ω; (b) $V = 2.4$ V; (c) $E = 0.079$ N/C; (d) $P = 60$ W.

7. (a) $I = 2.14$ A; (b) $\tau = 1.1210^{-6}$ s; (c) $t = 5.1610^{-6}$ s.

Chapter 26

COMMON PITFALLS

1. False. In this case the magnetic force is zero, so it cannot do any work. Generally, however, magnetic

forces are always perpendicular to the velocities of the particles they act on; thus they *never* do any work on the particles.

2. The wire is aligned with the magnetic field.

3. True. For this reason, the period of the orbit does not vary with either the speed or the radius of the orbit.

4. If the electron travels too slowly, the electric force will dominate and the electron will drift upward. If the electron moves too quickly, the magnetic force will dominate, causing the electron to drift downward.

5. True

6. If the forces do not have the same line of action, then the forces can combine to exert a net torque, causing the object to rotate, even though the center of mass does not accelerate.

7. False. Outside the magnet, the field lines appear to leave the north pole and end at the south pole but in actuality they do not. Magnetic field lines never begin or end; they are continuous through the magnet.

8. Charges build up on the sides of the Hall conductor only while there is a net force on the mobile charges. Once the induced electric force is equal in magnitude to the magnetic force, the charges stop accumulating. If there is no net force on the charges, they will go straight through the conductor— which is what happens when a charged particle with the proper speed goes through a velocity selector. The Hall voltage at which this occurs depends on the drift speed of the charges.

TRY IT YOURSELF–TAKING IT FURTHER

1. If the magnetic field were directed out of the page, then the current would have to be reversed. If the magnetic field were directed upward, then no amount of current would result in a zero net force on the wire because the magnetic force would always be directed perpendicular to the plane of the page.

2. If $\vec{B}$ were a function of y instead of x, then y would be the integration variable. This means you would have to find $y(x)$ instead of $x(y)$. Because you must integrate in the direction of the current, the lower limit of integration would be $L\sin\theta$, and your upper limit of integration would be zero.

3. The cyclotron frequency and potential difference between the dees would remain the same. With classical expressions, the maximum orbital radius is not even possible, as it would require speeds greater than that of light. So a different approach is clearly necessary, with additional physics knowledge.

4. The more massive negatively charged particle will have a larger radius, with a trajectory shown by the dotted line. The less massive positively charged particle will have a smaller radius of curvature and its path will curve in the opposite direction, as shown in the dashed-line path.

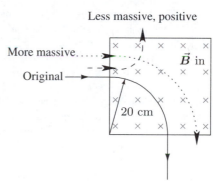

5. Potential energy is given by $U = -\vec{\mu} \cdot \vec{B} = \mu B \cos 135° = 15.0$ nJ.

6. The loop will initially rotate to point its dipole moment in the direction of the magnetic field. At this instant there is no net torque on the loop, but because of its rotational momentum it will continue to rotate past this point, slowing down. After it momentarily comes to rest, it will rotate back toward its initial position. If the system is frictionless, this harmonic oscillation will continue indefinitely.

7. If the magnetic field is out of the page, the forces on the charged particles are reversed, so the charge carriers must be positively charged. The drift velocity will remain the same.

QUIZ

1. True

2. True

3. The simplest way I can think of is to reverse the direction of the velocity of the particle. If the force is magnetic, then reversing the direction of the velocity will result in reversing the direction of the force. However, if the force is electric, then reversing the direction of the velocity will leave the force unaffected.

4. (a) The electric field is perpendicular to both the particle velocity and the magnetic field. Using the right-hand rule, we know that the magnetic force exerted on the positively charged particles moving to the right is toward us. (We are viewing the beam from the side.) The electric field must be away from

us if these particles are to pass undeflected. (b) Yes; reversing the direction of the velocity reverses the direction of the magnetic force, but not that of the electric force. Thus, the two forces are now in the same direction and the positively charged particles are deflected in the direction of the electric field. (c) Changing the sign of the charge results in both forces reversing direction. They still add to zero so there is no deflection.

5. Applying Newton's second law to a charged particle whose velocity is at right angles to a uniform magnetic field $\vec{B}$ results in the equation $qvB = mv^2/r$ (or $r = mv/(qB)$)—thus as long as the particles have the same charge, the radius of their orbits depends on their momentum, not their velocity alone.

6. $B = 0.511$ T

7. (a) $F = 3.03 10^{-3}$ N, (b) $F = 5.28 10^{-3}$ N

Chapter 27

COMMON PITFALLS

1. True. Imagine yourself, the observer, grabbing the bullet with your right hand with your thumb pointing away from you. Your fingers, as viewed by you, then curl clockwise. But because the charge on the bullet is negative, the magnetic field lines around the bullet are directed counterclockwise.

2. The electric field is given by $\vec{E} = kq\hat{r}/r^2$. Comparing this with the magnetic field expression, we first see a proportionality constant, either k or $\mu_0/(4\pi)$. We then see a source term, which for $\vec{E}$ it is simply q and for $\vec{B}$ is $q\vec{v}$. The source term is then multiplied in some way by $\hat{r}$, which has the same meaning in both expressions. Finally, both fields vary as $1/r^2$, where r is the distance from the point charge to the location where the field is measured.

3. False. The magnetic field on the surface varies as $\sin\theta$, where θ is the angle between the current element vector and the vector from the particle to the field point on the surface of the sphere. The angle θ ranges between zero and π.

4. Attractive magnetic forces between adjacent turns of the spring (think of them as parallel wires) cause the spring to compress longitudinally. The spring contracts and raises the mass. This force gets stronger as the turns get closer together, so the spring continues to collapse. This effect can be quite strong, and the construction of high-field laboratory solenoids has to be very rugged to keep them from imploding.

5. False. However, it is only for situations with sufficient symmetry that Ampre's law is very practical for determining the magnitude of the field.

6. The number of turns per unit length decreases as the coil is stretched, so the magnetic field inside, which is $B = \mu_0 n I = \mu_0 (N/L) I$, decreases. The magnetic dipole moment $\mu = NIA$ remains the same.

7. False. Because magnetization in a diamagnetic material is directed oppositely to the applied field, a diamagnetic material is repelled by either pole of a magnet.

8. When the field of the magnet is applied to the iron nail, the magnetization of the iron in the nail is very large. When the field is applied to the aluminum nail, the magnetization of the aluminum, which is paramagnetic, is many thousands of times smaller. The force of the non-uniform magnetic field of the magnet on the strong magnetic dipole moment of the iron is much greater than its force on the weak magnetic dipole moment of the aluminum.

TRY IT YOURSELF–TAKING IT FURTHER

1. In these situations $\vec{v}$ and $\hat{r}$ are parallel, and by definition the cross product of two parallel vectors is zero.

3. $\vec{B} = - \left(9.60 10^{-5}\ \text{T}\right)\hat{\imath}$. Because this point is the same distance from the sources as P, the magnitude of $\vec{B}$ will be the same. By the right-hand rule and symmetry, we can see that the z components will again cancel, but the resulting field will now be in the $-x$ direction.

4. Because P is located in the center of the $\frac{3}{4}$ circle, we know that $d\vec{\ell}$ is everywhere perpendicular to $\hat{r}$. In addition, we know that each $d\vec{\ell}$ is a distance r from P, and carries the same current I. This means each current element contributes identically to the magnetic field. The magnitude of the cross product simply becomes $d\ell$, and everything else is constant. When we integrate $d\ell$, we get a length equal to $\frac{3}{4}$ the circumference of a circle of radius r.

5. Ampre's law is certainly valid, but it is not useful in this case. Because the wire is short, there is not sufficient symmetry for Ampre's law to be properly applied in this instance.

6. Yes. If the outside conductor carried more current than the inside conductor then the sign of I_C will change outside the cable, reversing the direction of B_t.

7. Since iron is ferromagnetic with a much larger susceptibility, both the magnetization and net magnetic field inside the solenoid would be greatly increased.

QUIZ

1. True

2. True

3. Because the currents in the two loops are both clockwise, the two magnetic dipole moments are parallel. If you think of these loops as bar magnets, you can visualize that they're aligned with north pole next to north pole and south pole next to south pole, so they repel each other. In terms of the currents, the strongest force is between the two currents that are closest together. These are oppositely directed, and oppositely directed currents repel each other.

4. The ferromagnetic domains may get knocked out of alignment when a permanent magnet is jarred.

5. The two fields far from the dipoles look identical. The directions of both the magnetic dipole moment of the current loop (right-hand figure) and the electric dipole moment of the electric dipole (left-hand figure) are toward the right. The main difference is that the direction of the field "inside" the current loop is the same as the direction of the magnetic dipole moment (to the right), and the direction of the field inside the electric dipole is opposite to the direction of the electric dipole moment.

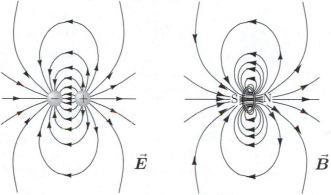

6. (a) $B = 0.0603$ T; (b) $I = 12.0$ A

7. $\vec{B} = 1.79 \times 10^{-3}$ T, into the page

Chapter 28

COMMON PITFALLS

1. True

2. As someone with a pacemaker or other electronic device walks through regions of spatially varying magnetic fields, the changing magnetic flux through the electronic circuitry will induce an emf in the device. This extra emf could cause unwanted, and possibly fatal, malfunctions of the device.

3. False. The induced emf would oppose not the magnetic field but the *change* in magnetic flux that induced the current.

4. A current is induced in loop b while the current in loop a is changing. If the current is in the direction shown and is increasing, the flux of its magnetic field $\vec{B}_a$ through loop b is upward and increasing. In accordance with Lenz's law, the direction of the induced current I_b in loop b is such that the flux of its magnetic field $\vec{B}_b$ through loop b is downward, opposing the change in flux that produced it. This means that I_b is antiparallel to I_a. Because the currents are antiparallel, the two loops repel. After I_a stops changing, $I_b = 0$ and there is no force between the loops.

5. True. As long as the magnetic field has some component perpendicular to the plane of the circuit, there will be a changing flux, resulting in an induced emf, and an induced current. As a result, the sliding wire will feel a magnetic drag force, which must be compensated for.

6. The induced emf will create an induced current in the clockwise direction, which will exert a force to the left on the sliding wire. This will gradually slow the wire until it comes to a stop.

7. True in most cases. Even though the circuits are not in physical contact, each is affected by the change in flux of the magnetic field due to the current in the other circuit. (For certain very specific geometries this change in flux is zero.)

8. Inductance is flux divided by current. Since flux is proportional to current, the current terms cancel, leaving only geometric properties.

9. True

10. The most enlightening forms of these expressions are $U_{\mathrm{m}} = \frac{1}{2}LI^2$ and $U_{\mathrm{E}} = Q^2/(2C)$. We see that both energies are proportional to the square of a quantity related to charge (either I^2 or Q^2). However,

the magnetic energy stored is proportional to the inductance, but the electric energy stored is inversely proportional to the capacitance.

11. False. If the resistance is negligible, then the time constant L/R is very large for a finite inductance L.

12. The current drops suddenly toward zero when the switch is opened. This current drop induces a very large emf in the inductor, which produces a large potential difference across the switch gap. This potential is large enough to cause dielectric breakdown of the air in the gap; this causes the spark, which allows current to continue to flow for a brief period.

TRY IT YOURSELF–TAKING IT FURTHER

1. For the semicircle with $x > 0$, $\phi_{\mathrm{m}} = 0.0384$ Wb, because the flux is all of the same sign. For $y < 0$, $\phi_{\mathrm{m}} = 0$ again, because the flux is positive for $x > 0$ and equal in magnitude but negative for $x < 0$.

2.

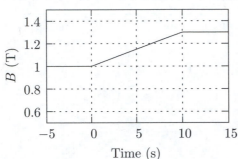

3. Zero. Charge travels in one direction for the first half of the rotation, but then the change in flux reverses directions, so the same amount of charge flows in the opposite direction for the second half of the rotation, leading to no net movement of charge for the full rotation.

4.

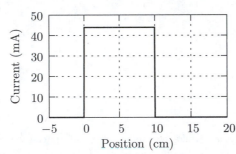

5. The current will be counterclockwise as viewed. The force on the left-hand segment will be zero because there is no magnetic field there. The force on the top and bottom segments will be downward and upward, respectively. The force on the right-hand segment will be to the left.

6. The magnetic force on the tether will slow down the shuttle, which will cause its orbital radius to decay in the absence of any additional thrust.

7. The outside edge is at a higher potential. A positive charge moving in the counterclockwise direction will feel a force toward the outside edge, and negative charges will feel a force toward the center.

8. In the new situation, it is much more difficult to calculate the flux through the solenoid from the coil because the magnitude and direction of $\vec{B}$ change along the length and radius of the solenoid, respectively. The problem is not impossible, though. From the existing problem we can calculate the mutual inductance M. Once we have that, the problem is straightforward: $\mathcal{E}_{\mathrm{solenoid}} = M\, dI_{\mathrm{coil}}/dt$.

9. 5.0310^{-7} C. The charge that passes through is proportional to the total magnetic field. So the charge without the magnetic material is the charge with the magnetic material multiplied by the ratio of B_{app} to the magnetic field with the material.

10. L is proportional to A, so it is doubled. The energy stored is proportional to L, so it is also doubled.

12. If $I < 5$ A, then $V_b < V_a$, which would mean the potential difference across the inductor is opposing that of the battery, which means the current is increasing.

QUIZ

1. True

2. True. Any abrupt change in current produces a large back emf opposing the change.

3. The falling bar magnet induces a current in the wall of the pipe. In accordance with Lenz's law, the direction of this induced current is such that it produces a magnetic field that exerts forces on the

magnet that oppose its motion. The speed of the magnet therefore increases less rapidly than in free fall until it reaches a terminal speed at which the magnitudes of the gravitational and magnetic forces acting on the magnet are equal.

4. As the conducting bar is moved to the right, a downward magnetic force $-e\vec{v}\vec{B}$ acts on each and every free electron in the rod. This causes the electrons to drift downward (down the page), making the lower end of the rod negative and the upper end positive. This charge distribution produces a downward electric field in the rod. After the free charges have stabilized, no current (moving charge) exists in the bar, and so there is no net magnetic force on the bar opposing its motion. Therefore, no external force is needed to keep the bar moving at constant velocity.

5. The electric field is nonconservative. The electric field "circulates" around the area through which the flux changes, and every time a charge makes a full round trip, its potential increases further, rather than having $\Delta V = 0$ for a round trip which is required for a conservative field.

6. $I = \mathcal{E}/(2R)$; $dI/dt = \mathcal{E}/(2L)$

7. $\mathcal{E} = 4.21$ mV

Chapter 29

COMMON PITFALLS

1. True

2. True

3. It means that the current peaks one-fourth of a period after the voltage drop peaks. One-fourth of a period corresponds to a phase difference of 90°.

4. True

5. The statement is false. The average power delivered by an ideal transformer equals the average power it receives. Transformers are not power sources, so they cannot deliver more average power than they receive. In fact, they deliver less average power than they receive because of internal dissipative processes.

6. The electrical energy stored in the capacitor is $U_e = Q^2/(2C)$, which varies as $\cos^2(\omega t)$, and the magnetic energy stored in the inductor is $U_m = LI^2/2$, which varies as $\sin^2(\omega t)$. Therefore when a maximum amount of energy is stored in the capacitor, no energy is stored in the inductor, and vice versa.

7. True. The instantaneous voltage drops are the same because they are in parallel.

TRY IT YOURSELF–TAKING IT FURTHER

1. 120 V. A much smaller proportion of the total power is dissipated in the cord, which means more is delivered to your device. It is much more efficient than the 12.0-V delivery system.

2. The average power delivered to the inductor and capacitor is zero, because the current and voltage are 90° out of phase. The average power delivered to the resistor is $I_{\text{rms}}^2 R = 16.0$ W.

3. The rms current will increase. As ω is increased, X_C is reduced, so I_{rms} will increase.

6. (a) The maximum voltages across the capacitor, inductor, and resistor are not in phase with each other, so their sum can be larger than the maximum applied emf. (b) At resonance, the voltage across these elements must cancel each other, so their maximum voltages must be the same.

QUIZ

1. True

2. True

3. It saves copper. The primary power loss in transmission equals $I_{\text{rms}}^2 R$, where R is the resistance of the transmission lines, and the rate at which energy is transported equals $V_{\text{rms}} I_{\text{rms}}$. Thus, at high voltage, power can be transmitted with a lower current, and the lower the current, the less the transmission loss. The other way to reduce transmission loss would be to decrease the resistance of the transmission lines. The only practical way to do that is to use thicker wire, which requires more of the metal the wires are made of. The voltages are stepped down for consumer safety.

4. An inductor with resistance is equivalent to a series LR combination. The tangent of the phase angle equals the ratio of the inductive reactance to the resistance. The phase angle varies with frequency because the inductive reactance does.

5. The instantaneous current in each device is the same because they are in series. The voltage drop across the inductor leads the current by 90°, whereas the voltage drop across the capacitor lags the current by 90°; thus, the two voltage drops are 180° out of phase. The equivalent reactance of the combination is the ratio of the rms voltage drop across the combination to the rms current through it. Because the two voltage drops are 180° out of phase, the rms voltage drop across the combination equals the difference of the individual rms voltage drops; so the equivalent reactance equals the difference

between the individual reactances. Represented as a phasor, the voltage across the combination is $\vec{V} = \vec{V}_L + \vec{V}_C$. Because $\vec{V}_L$ and $\vec{V}_C$ are oppositely directed, the magnitude of $\vec{V}$ equals the difference between their magnitudes.

6. (a) $R = 65.0 \ \Omega$; (b) $X_{\text{net}} = 94.9 \ \Omega$
7. (a) $f_{\text{res}} = 225$ Hz; (b) $P_{\text{av}} = 3.13$ W

Chapter 30

COMMON PITFALLS

1. True
2. A closed curve can bound many different surfaces, including nonplanar surfaces. In situations where the current is discontinuous, for example when charging a capacitor, it is possible for the current to penetrate one of the surfaces bounded by a closed curve, but not another surface. This would result in two different values for the magnetic field along the curve, depending on which surface was chosen. The idea of displacement current removes this ambiguity.
3. True
4. The electric field produced by a time-varying magnetic field induces currents that are strong enough to detect easily—for example, a transformer can be used to light a light bulb. The magnetic fields produced by time-varying electric fields are weak and, therefore, detection requires more sensitive equipment.
5. True
6. The radiation pressure $P_{\text{r}} = I/c$ is the momentum per unit time per unit area carried by a wave. On a perfectly reflecting surface both the incident wave and the reflected wave carry momentum and both exert pressure on the surface. Thus the net pressure is twice the pressure exerted by a wave incident on a perfectly absorbing surface.

TRY IT YOURSELF–TAKING IT FURTHER

1. The problem would be significantly easier because no physical current would penetrate the surface.
2. Yes. The flux through the larger-radius loop will be larger by the square of the radius, but the circumference is proportional to the radius.
4. The magnitude of the voltage can be increased by using a longer antenna or moving closer to the source. There are also resonant effects, which occur when the length of the antenna is some multiple of the wavelength of the radiation. Discussion of these resonant effects, however, is beyond the scope of an introductory text.
5. The area can be increased, the antenna can be moved closer to the source, or the frequency of the transmitted signal can be increased. Most commonly the effective area is increased by adding multiple turns.

QUIZ

1. False. The displacement current through a surface is the electric permittivity of free space times the rate of change of the flux of the electric field through the surface. It does not represent the rate of flow of free charges.
2. False. Maxwell's equations apply to all $\vec{E}$ and $\vec{B}$ fields.
3. The equation $\oint \vec{B} \cdot d\vec{A} = 0$, or Gauss's law for magnetics, denies the existence of magnetic monopoles. The net flux through a closed surface would be nonzero if a magnetic monopole existed inside the surface.
4. The loop antenna works by induction. The changing magnetic field induces an emf and, therefore, a current in the loop. Thus the loop must be oriented such that the magnetic field of the wave produces a changing magnetic flux through the loop, preferably with the magnetic field perpendicular to the plane of the loop. This means that the plane of the loop should contain the electric dipole moment of the broadcast antenna.
5. The direction of the Poynting vector is the direction of propagation of the wave: $\vec{S} = \vec{E}\vec{B}/\mu_0$.
6. 1.11 m^2. A person presents a total area less than 1 m^2 to the sun and people are not 100 percent reflective. As such, it is unlikely we can feel the radiation pressure which would be less than the weight of a mosquito distributed over an entire body.
7. $P = bI^2L$ =power dissipated by the resistance of the wire

Chapter 31

COMMON PITFALLS

1. False. Only about 4 percent of the incident intensity is reflected.

2. The sketch shows a beam of light in a medium, like the atmosphere, in which the refractive index decreases as you go upward. The light in the lower portion of the wavefronts travels less rapidly than the light in the upper portion. In this situation the wavefronts in the beam turn downward as shown. When the Sun is just above the horizon, the light from the bottom edge of the Sun is bent downward to a greater degree than is the light from the upper edge of the Sun. Thus the Sun appears flattened when it is near the horizon.

3. False. Only the reflected light is completely polarized.
4. Most of the glare you want to reduce is reflection off horizontal surfaces—from the sun off a highway or the surface of a swimming pool, for instance. This reflected light is partially polarized in a horizontal plane. To block this glare you want the transmission axis of the polarizing lenses to be vertical.

TRY IT YOURSELF–TAKING IT FURTHER

1. The rate of rotation is quite high, so you might suspect that Foucault's rotating mirror had more than eight surfaces.
2. The total round-trip delay is a minimum of 8.66 minutes—that is a lot of time for something to happen. The rovers should be able to react at least somewhat to their surroundings on their own, with only occasional input from Earth, in order for those reactions to occur in a timely fashion.
3. $I_{reflected} = 0.111I_0$ because the total amount of light must be conserved.
4. You would have to change in index of refraction of water so that it was smaller than the index of refraction of air. Under that condition total internal reflection cannot occur.
5. $\theta_p = 22.3°$. Yes;, $\theta_p < \theta_c$ because the arctan of some value is always less than the arcsin of that same value.
6. The transmitted intensity is a function of θ, the angle of the transmission axis of the middle polarizer. Taking the derivative of the intensity with respect to θ and setting it equal to zero yields, among other solutions, $\theta = 0°$, $45°$, and $90°$. The first and last of these correspond to intensity minima. $45°$ is the angle of maximum transmission.

QUIZ

1. True
2. True
3. The frequency of a light wave in any medium is the number of crests that pass any given point per unit time. When this wave is incident on a boundary, the rate at which the crests arrive at any point, including points on the boundary, is equal to the frequency in the first medium. The rate at which these crests leave the boundary and enter the second medium is the frequency in the second medium. Because the crests do not pile up at the boundary, the two frequencies must be equal. The index of refraction is greater in the second medium, so the crests travel more slowly than they did in the first medium. Because they leave the boundary at the same rate (frequency) at which they arrive, the crests are closer together in the second medium; they have to be because they have a shorter distance to travel in the time between the arrivals of successive crests at the boundary. The distance separating adjacent crests is the wavelength, which is the product of the speed of the wave and the time between the arrival of the crests. Hence the wavelength is shorter in the medium in which the wave travels more slowly.
4. Reflection from a dry highway is diffuse; thus some of the light from your headlamps is reflected back at you, enabling you to see the road surface. When it rains, the water surface reflects the light specularly, so most of the headlamp beam is reflected forward and little or none of the light gets back to your eye.
5. Because of the refraction at the air–water boundary, you see the underwater portion of the oar at a different position from where it actually is, and so the shaft of the oar appears bent at the water's surface.
6. $40.5°$
7. (b) 0.333 cm

Chapter 32

1. False. Paraxial rays are rays that remain near enough to the axis (but not necessarily parallel to it), that we can use small-angle approximations to analyze their reflection and refraction.
2. Actually, a plane mirror does neither, but inverts objects back to front as illustrated. If the mirror inverted right and left, then the object's right hand that points east would appear on the image as a right hand pointing toward the west. Because the image is inverted back to front, the object who faces north is transformed into an image that faces south. Also, the object's right hand is transformed into a left hand in the image.

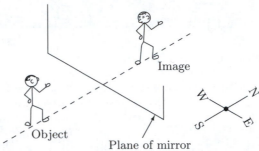

3. False. It is half the radius of curvature.
4. The focal length of a lens is proportional to the difference in index of refraction between the lens material and its surroundings. Thus if the lens material has a refractive index greater than that of air but less than that of the fluid, its focal length would change sign when you immerse it.
5. True
6. The aging eye loses its ability to accommodate—that is, the eye's lens is not flexible enough to change the focal length of the cornea-lens system by very much. Bifocals compensate for the eye's poor accommodation by giving it the choice of two corrective lenses of slightly different focal length. (Some of us decrepit types need trifocals!)

TRY IT YOURSELF–TAKING IT FURTHER

1. Following the same procedure that was laid out in this problem, but for the south mirror, results in a separation distance of 4.00 m.
2. Using the mirror equation, we see that s' will be negative, and the resulting image virtual, if the object position s is less than the focal length f of the concave mirror.
3.

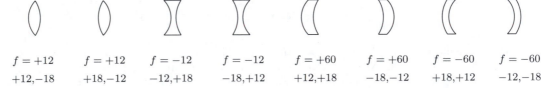

| $f = +12$ | $f = +12$ | $f = -12$ | $f = -12$ | $f = +60$ | $f = +60$ | $f = -60$ | $f = -60$ |
| $+12,-18$ | $+18,-12$ | $-12,+18$ | $-18,+12$ | $+12,+18$ | $-18,-12$ | $+18,+12$ | $-12,-18$ |

4. Because the image distance is negative, the image is formed to the left of the lens, and is a virtual image. To determine the magnification, you need to apply the magnification expression twice, once for each object. The image is inverted, and three times larger than the original object. To see the image, a viewer would have to look to the left through both lenses.
5. In the limit that we consider the lens of the glasses to be "very close" to the eye, $M = (1.10 \text{ m})P = 1.40$.
6. Use the thin-lens equation to find the location of the image. Use the object and image distances to find the magnification. Use the magnification to find the diameter of the image, which is 216 μm.

QUIZ

1. True. The angle of reflection equals the angle of incidence for all wavelengths (colors). Thus chromatic aberration does not occur in reflective optical elements (mirrors).
2. False. Only a real image can be projected on a screen. In the case of a virtual image, the light you see only *appears* to diverge from the image.
3. Additional information is needed. In accordance with the lens-maker's equation, if the radius of curvature of the front surface is the larger, it is a diverging lens; if the radius of curvature of the front surface is the smaller, it is a converging lens.
4. The film is placed at the position of the real image formed by the objective lens (or mirror), and an eyepiece is not used.
5. The mirror has to be half your height, as illustrated. Perhaps surprisingly, the answer does not depend

on how far from the mirror you stand.

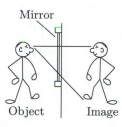

Mirror

Object | Image

6. $n = 1.524$
7. (a) $f_e = 3.13$ cm; (b) $s = 0.832$ cm

Chapter 33

COMMON PITFALLS

1. True if the light is incident normal to the interface.
2. Because the wavelength of visible light is so small, the diffraction and interference of light are usually too small to see without coherent optical sources and very small apertures which were not available to Newton.
3. False. One must consider phase shifts before deciding. It would be destructive if 180° phase shifts occur either at the first or the second surface but not at both. Otherwise, it would be constructive.
4. The colored bands are due to the interference of light reflected at the front and back surfaces of the thin film. The different colors are due to the varying thickness of the film. At one point on the film, for instance, there is an interference maximum in the blue reflected light and a minimum in the red light, so the film looks blue. At another point where the film is slightly thicker, it may preferentially reflect red light. Thus when we look at sunlight reflecting from an oil slick, we see a shifting rainbow of colors in the reflected light.
5. True. To cancel completely, the waves must have the same amplitude. Because for a given frequency the intensity of a wave is proportional to the square of its amplitude, coherent waves having the same amplitude also have the same intensity.
6. The power passing through two narrow slits is twice that passing through one narrow slit alone, though the distribution of the power over the observation screen is very different. With one slit the entire screen is almost uniformly illuminated. This is not so in the two-slit pattern. In the two-slit pattern, the area of the screen that is brightly illuminated is approximately equal to the area that is dimly illuminated. A large intensity over a small area can deliver the same power as a low intensity over a large area. To produce four times the intensity means that twice the power is delivered to an area one-half as large.
7. False. Diffraction is important for both large and small apertures. For relatively small apertures, the resulting diffraction pattern is readily observed, and the resulting "shadow" is much different from the expected. As you will see in section 33-7, for relatively large apertures like those found in telescopes, diffraction often limits the resolution of the instrument.
8. The first minimum in the single-slit diffraction pattern occurs at $\sin\theta = \lambda/a$, where a is the width of the slit. Because $\sin\theta$ cannot be greater than 1, this minimum can be observed only if $\lambda < a$. Thus the maximum wavelength for which the first minimum in the single-slit diffraction can be observed is $\lambda_{\max} = a$. For wavelengths very much smaller than a, diffraction effects become unobservably small, and all you see is the geometric shadow of the slit. To say that diffraction effects become unobservably small is to say that the diffraction maxima and minima are so closely spaced that the eye cannot resolve them. The wavelength for which this spacing is just resolvable is the minimum wavelength for which the diffraction pattern can be observed.
9. False. The principal maxima occur at the same angles, but are sharper (narrower and more intense).

10.

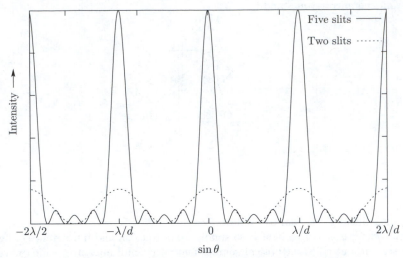

11. True. The usual expression of this is Rayleigh's criterion, $\alpha_c \approx 1.22\lambda/D$.

12. Since $\frac{1}{4}$ arc-sec is 1.2110^{-6} rad, Rayleigh's criterion is $1.22\lambda/D = 1.2110^{-6}$. For wavelengths of visible light, this corresponds to a diameter D of 55 or 60 cm (about 2 ft). Building a telescope with a diameter much larger than this won't improve the resolution significantly as long as you have to look through Earth's atmosphere. The main reason optical telescopes bigger than this are built is for greater light-collecting power, not resolution. In addition, telescopes with adaptive optics can now compensate for atmospheric variations, greatly improving the possible viewing resolution.

13. True. The larger the grating, the greater the number of lines that can be illuminated.

14. Both interference and diffraction are basic to the operation of a grating. The pattern produced by the grating is just a very-many-slit interference pattern; on the other hand, the wide-angle pattern wouldn't exist if it weren't for the diffraction of the light at each slit.

TRY IT YOURSELF–TAKING IT FURTHER

1. For transmission, consider one path that goes straight through the film, and another path that is reflected first off the film–glass interface, travels back up through the film, is reflected off the film–air interface, and travels back through the film a second time. There are no phase changes on reflection for either path. The first path experiences no reflections, and the second path is always reflected off a medium with a lower index of reflection. So the phase difference is due entirely to the path difference. The condition for destructive interference then becomes $2tn_{\text{film}}/\lambda = m + \frac{1}{2}$. The first few values of λ that are not present in the transmitted light are 3.96 μm, 1.32 μm, 791 nm, 565 nm, and 437 nm.

2. The fringes will be spaced closer together, because less change in thickness is required to move from one interference maximum to the next.

3. When the plastic is put in front of one of the slits, there will be a phase difference between the two beams leaving the slits, as shown. The light traveling through air will experience some phase change as it travels the thickness Δr through the air, but the light traveling through the plastic will experience a different phase change as it travels the distance Δr through the plastic. This extra difference in phase can be calculated to be $\pi/2$. The result is a shift in the interference pattern. The spacing is still 4 cm, but the maxima have moved along the y axis.

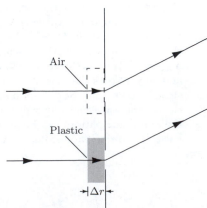

4. The third minimum will disappear next. As the slit widths are increased, the spreading of the beam is decreased, so the diffraction minimum will occur at a smaller angle.

5. The central maximum will be wider. Relative to the wavelength, the slit will be reduced in size, thus increasing the spread of the beam.

6. To find the separation of the images of the two sources, simply use the fact that the object and image

must both subtend the same angle α_c, so the ratio of object or image separation to the distance from the lens must be equal. The image will be formed at a distance equal to the focal length of the lens. $d' = 2.91 \ \mu$m.

7. 1.00 arc-sec corresponds to a separation of 96.9 cm on Earth. Since the resolution with atmospheric turbulence is less than the diffraction limited resolution, the atmospheric resolution is the limiting factor, so the telescope can resolve only objects separated by about 1 m.

8. The resolving power of the grating relates the number of lines illuminated to the wavelength resolution. This relation can be used to find the number of lines that must be illuminated. Then, from the line spacing, we can determine the required beam width, $d = 289 \ \mu$m.

QUIZ

1. False. The phase differences affect the locations of the maxima and minima, but interference occurs whenever the two sources are coherent and light waves from the two sources superpose.

2. True. Light is diffracted at each slit, and light from one slit interferes with the light from the other slit.

3. AR coatings are placed on the front surface of the glass. The index of refraction of the coating is greater than the index of refraction of air and less than the index of refraction of glass. This means that 180° phase shifts are associated with the reflections at both surfaces. Also, the thickness of the coating is one-fourth the wavelength in the coating. Thus the light reflecting at the back surface has a path length that is one-half a wavelength longer than the light reflected at the front surface. Consequently, the reflected light waves are 180° out of phase and interfere destructively.

4. The wavelength of an electron is inversely proportional to its momentum. Electron microscopes use electrons with wavelengths that are much smaller than those of light, so, in accordance with Rayleigh's criterion, much smaller objects can be resolved distinctly by an electron microscope.

5. The two images of the actual light source in the two plane mirrors act as two point sources as seen below. The two image "sources" are not in phase because they are at different distances from the source; but they are coherent because all the light actually comes from a single source.

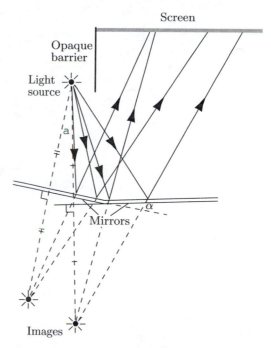

6. 4 mm

7. 902 nm